Faith and Language Practices in Digital Spaces

Full details of all our publications can be found on http://www.multilingual-matters.com, or by writing to Multilingual Matters, St Nicholas House, 31-34 High Street, Bristol BS1 2AW, UK.

Faith and Language Practices in Digital Spaces

Edited by
Andrey Rosowsky

MULTILINGUAL MATTERS
Bristol • Blue Ridge Summit

*The volume is dedicated to the memories
of Joshua A. Fishman (1926–2015)
and Tope Omoniyi (1956–2017)*

DOI 10.21832/ROSOWS9276
Library of Congress Cataloging in Publication Data
A catalog record for this book is available from the Library of Congress.
Names: Rosowsky, Andrey, 1956- editor.
Title: Faith and Language Practices in Digital Spaces/Edited by Andrey
  Rosowsky.
Description: Bristol, UK; Blue Ridge Summit, PA: Multilingual Matters,
  [2018] | Includes bibliographical references and index.
Identifiers: LCCN 2017029081| ISBN 9781783099276 (hardcover: acid-free
  paper) | ISBN 9781783099283 (pdf) | ISBN 9781783099290 (epub) |
  ISBN 9781783099306 (Kindle)
Subjects: LCSH: Language and languages–Religious aspects. |
  Communication–Religious aspects. | Digital media–Religious aspects. |
  Social media–Religious aspects. | Religion–Computer network resources. |
  Information technology–Religious aspects.
Classification: LCC P53.76 .F24 2018 | DDC 401/.47–dc23 LC record available at
  https://lccn.loc.gov/2017029081

British Library Cataloguing in Publication Data
A catalogue entry for this book is available from the British Library.

ISBN-13: 978-1-78309-927-6 (hbk)

Multilingual Matters
UK: St Nicholas House, 31-34 High Street, Bristol BS1 2AW, UK.
USA: NBN, Blue Ridge Summit, PA, USA.

The policy of Multilingual Matters/Channel View Publications is to use papers that
are natural, renewable and recyclable products, made from wood grown in sustainable
forests. In the manufacturing process of our books, and to further support our policy,
preference is given to printers that have FSC and PEFC Chain of Custody certification.
The FSC and/or PEFC logos will appear on those books where full certification has been
granted to the printer concerned.

Typeset by Deanta Global Publishing Services Limited.
Printed and bound in the UK by the CPI Books Group Ltd.
Printed and bound in the US by Edwards Brothers Malloy, Inc.

Contents

Part 3: Faith, Language and Online Televangelism

Part 4: Faith, Language and Online Ritual

Part 5: Afterword

Acknowledgements

The editor would like to thank each contributor for their chapter and cooperation. He thanks them in particular for their patience and forbearance during the editing process. He would like to thank the anonymous reviewers for their wise and practical advice which has significantly helped improve the final outcome. He would also like to express his and his fellow contributors' appreciation to the publishers for their encouragement, guidance and support during publication. When this manuscript had already been sent to the publishers, we received the sad news that one of its contributing authors and, more importantly, one of its main inspirations, Tope Omoniyi, had passed away. This volume is therefore dedicated to his memory as well as to that of his mentor, Joshua Fishman.

Figures

Tables

Contributors

Iyabode Deborah Akande is an early career scholar who is completing a doctorate in the Department of Linguistics and African Languages, Obafemi Awolowo University (OAU), Ile-Ife, Nigeria. She earned a bachelor's degree in the Department of African Languages and Literature in 1995 at the University of Lagos, Akoka, Nigeria, and has two masters degrees from OAU, Ile-Ife: one in Special Education and Curriculum Studies in 2005 and the other in 2013 in Yoruba with a bias in Oral Literature and Culture. Iyabode has authored two papers and a third one is coming out in June 2017 in the *Journal of Pan African Studies.*

Shaimaa El Naggar gained her PhD in Linguistics from Lancaster University in 2016. In her PhD, she explored the interaction between language, new media and religion. She revealed an original analytical framework for analysing the discourse of three case studies of American Muslim televangelists drawing on critical discourse analysis, multimodality and stylistics. Her research interests include sociolinguistics, critical discourse studies, stylistics, new literacy studies, multilingualism and cultural studies. While undertaking her PhD studies at Lancaster University, she co-convened the research group 'Language, Ideology and Power' together with Distinguished Professor Ruth Wodak.

Tope Omoniyi (1956-2017) was Professor of Sociolinguistics in the Department of Media, Culture and Language at the University of Roehampton. He was, with Joshua Fishman, one of the two founding members of the Sociology of Language and Religion (SLR) international research network. He was a dear friend and colleague to all the contributing authors in this volume and will be greatly missed.

Rajeshwari V. Pandharipande is Professor Emerita of Linguistics, Religion, Sanskrit, Comparative Literature, Asian American Studies and Campus Honors Program at the University of Illinois at Urbana-Champaign (UIUC) and a permanent member of the South Asia Committee, University of Chicago. Professor Pandharipande holds two PhD degrees – one in Sanskrit

Literature, Religion and Philosophy from Nagpur University, India, and the other in Linguistics from the University of Illinois at Urbana-Champaign, USA. The primary focus of her research and teaching has been South Asian languages (syntax, sociolinguistics and literature), Asian mythology, sociolinguistics, the language of religion and Hinduism in India and in the diaspora. Her publications include five books, 80 research papers/book chapters and three edited books. In 2008, she received the prestigious award with the permanent title 'University Distinguished Teacher Scholar' for her outstanding record of teaching and research at UIUC.

Andrey Rosowsky is a Senior Lecturer in the School of Education at the University of Sheffield. His research explores literacy and language practices of faith-based supplementary schools. He has an interest in Qur'anic literacy and its relationship to notions of performance and how poetry and song in heritage languages and English are instrumental in reviving both religious and linguistic practices. He currently leads an AHRC international interdisciplinary research project on performance and religious practice entitled 'Heavenly Acts: Aspects of Performance through an Interdisciplinary Lens'. He is the author of *Heavenly Readings: Liturgical Literacy in a Multilingual Context* (Multilingual Matters, 2008).

Oladipo Salami (DPhil) is Professor of Language and Linguistics in the Department of English, Obafemi Awolowo University, Ile-Ife, Nigeria. He had his undergraduate education at the University of Ibadan, Master of Arts (Linguistics) from Obafemi Awolowo University, Nigeria, and a DPhil from the University of Sussex, Brighton, UK. Professor Salami has extensive university teaching and research experience spanning over three decades. His areas of teaching and research interest include phonetics and phonology; sociolinguistics; psycholinguistics; language policy; the English Language in Nigeria; and language and religion. He is a member of the international networks on the Sociology of Language and Religion (SLR) and Heavenly Acts.

Thor Sawin is an Assistant Professor of Applied Linguistics at the Middlebury Institute for International Studies at Monterey, where he has taught in the Teaching Foreign Language and TESOL programs since earning his doctorate in Linguistics in 2013. He is interested in acquired second languages as resources for identity mediation in a range of multilingual environments – from faith communities to social media postings – and the mechanisms of acquiring those languages in field settings, especially by personnel of transnational organizations. He has presented at dozens of international conferences, and is published in *Crossroads of Language, Interaction and Culture* and *Reconsidering Development*.

Tatjana Soldat-Jaffe is an Associate Professor of Linguistics in the Department of Modern Languages and Linguistics at Florida State University. Her research focuses on language identity and minority languages. She is the author of *Twenty-First Century Yiddishism: Language, Identity, and the New Jewish Studies*, 'Yiddish without Yiddishism: Tacit Language Planning among Haredi Jews', 'The European Charter for Regional or Minority Languages: A Magnum Opus or an Incomplete Modus Vivendi?' 'Zombie Linguistics' and 'Translation: An Exercise in Midrashic Reading, or Translating the Intentio'.

Ana Souza is a Senior Lecturer in TESOL and Applied Linguistics at Oxford Brookes University, UK. She graduated in Language Teaching and Translation (Portuguese/English) in Brazil, did her MA in ELT at Thames Valley University (present West London University) and her PhD in Sociolinguistics at the University of Southampton. She completed her PGCE in Higher Education at Goldsmiths, University of London and became a Fellow of the Higher Education Academy. Her research interests include bilingualism, language negotiations, complementary schools, language planning (families/migrant churches), the teaching of Portuguese as a Heritage Language and training of language teachers (souzaana.wordpress.com).

Bernard Spolsky was born and educated in New Zealand, earned a doctorate from the Université de Montréal and taught in New Zealand, Australia, England and the United States before his appointment at Bar-Ilan University in Israel from which he retired in 2000 as Professor Emeritus. He has written 11 monographs, edited 24 books and published 245 articles and chapters in learned journals and books. A past president of International TESOL and also of the International Language Testing Association, he was awarded a Guggenheim fellowship, an honorary D.Litt. from Victoria University, and is a Fellow of the Linguistic Society of America.

1 Faith and Language Practices in Digital Spaces: An Introduction

Andrey Rosowsky

This is a challenging book. And writing its introduction is a challenging task in a number of ways. Firstly, this edited volume challenges the reader to make sense of a wide range of theory, context and data from different academic fields that, while diverse and in themselves complex, are nevertheless constrained within the nexus of faith, technology and language. Unlike some interdisciplinary work, which might lead the researcher to wherever the data takes him or her, there is here a self-imposed limitation to explore and account for specifically faith, linguistic and technological practices in their intricate and intimate relationship with one another. Such a limitation is imposed, of course, in the spirit of practicality, cohesion and realism. Any two of these areas would provide a lifetime's wealth of study for many of us. The addition of a third – which is not simply a context but a dynamic and independent academic path in itself – poses a serious challenge in coming up with coherent, meaningful and relevant insights and conclusions. It is, though, a necessary framework for making sense of this particular nexus within the complex, stochastic and fluid set of social processes which characterise the present age.

Furthermore, many of the chapters in this volume owe part of their origin to the pioneering work done by others in the field of the sociology of language and religion (SLR), a fledgling discipline which also tries to marry two very independent and historically recognised academic fields. Looking over our shoulders throughout this book, and I am sure of his presence now overlooking my computer as I write, is the sociolinguistic figure and legacy of the late Joshua Fishman (1927–2015), who throughout his life, but particularly towards the end of it, never ceased questioning the always giving connection between language and religion (Fishman & Omoniyi, 2006). More on this later.

Secondly, this book has been a challenge for its contributors and its editor, as despite the self-imposed constraints provided by our tridental focus, or because of them, a very wide range of contexts have been

1

identified as having something important to say about the way that faith and language interrelate through technology, particularly in this case, new technologies. A point that needs to be made from the start is that our conceptualisations of language and religion are broad and multifaceted. This is justifiable from a couple of perspectives. Firstly, the collection of authors in this volume draws on a number of different, but related, fields of research, each of which approaches the study of language and/or religion from a certain direction. The second justification is the inherent difficulty in coming up with satisfactory working definitions that encompass all the research included. In respect of religion, Fishman (2006: 14) famously said 'I am not prepared to define "religion" per se, accepting that the behaviors, beliefs and values that are deigned to be religious are more diverse than any of us are currently aware of'. The most one can say about 'language' in the context of this volume, that is consistently valid and meaningful, is that all contributors, whatever their discipline, treat language as a social practice.

This volume attempts to maintain a focus as much as possible on the three discrete areas of faith, language and digital (or online) spaces. The latter phrase I use here as a catch-all term to denote online communications and interactions either via computers, handheld devices or other forms of mobile technology. All three of these areas already have their 'sociologies' albeit the latter of more recent origin than the other two. The first, religion, has antecedents reaching back to Durkheim and Weber, while the second, language, it could be argued has its origins in the work of Saussure who made the important academic distinction between language in theory (*langue*) and language in practice (*parole*), though strictly speaking, the fundamentals of the sociology of language were established in the 1960s with the work of Labov, Fishman and Hymes. Thirdly, as befitting such a rapidly changing field of inquiry, it is easier to refer to a journal rather than any one scholar for research into new technologies and communication. The *Journal of Computer-Mediated Communication* has been at the forefront of this field for a number of years and the *Journal of Sociolinguistics* has had at least one issue dedicated to language and online spaces (10/4, 2006). While mentioning journals, it is also worth noting the existence of an online journal specifically focusing on online religion/religion online, the *Heidelberg Journal of Religions on the Internet*, which has been around since 2005 and is now into its ninth volume. The present volume, then, differs from these other approaches in seeking to submit all three of these areas, in combination, to scholarly gaze. It takes its place, therefore, within these three fields and offers a forum for their interaction and mutual effects.

The Sociology of Language and Religion

In Fishman's (2006) Decalogue, which outlines the principles underpinning any future SLR, a number of its precepts are relevant to a volume of this nature. His first precept is that a 'language (or variety) of religion always functions within a larger multilingual/multivarietal repertoire' (Fishman, 2006: 14), which speaks clearly to the findings of a number of authors here in that digital spaces, either technologically determined or, at least, influenced or shaped, encourage newer or transformed varieties of language which are thus added to a community's or individual's repertoire. Of course, evidence has to be presented that (a) such digital spaces are responsible in some way for these new or transformed varieties and (b) they are inflected to some extent with a faith dimension either missing from their offline equivalent contexts or, at least, consolidating them in some special or enhanced way. Either of these possibilities would be worth exploring within the remit of SLR.

The Decalogue's second precept also builds on the initial precept by claiming that such variation, where it occurs, takes place inter- and intra-societally. These varieties emerge on the basis of faith practices both within the same faith, broadly understood, and between faiths, equally broadly understood. For example, across a range of Christian contexts, obvious different varieties of English are spoken and performed in church practices of different denominations and even within single denominations. The same is true for other faiths and other traditions. Therefore, within a nominally monolingual and monofaith set of Christian practices, we may find a wide range of faith-inflected language varieties, vernacular and performative, such as in African American Vernacular English within churches belonging to the National Baptist Convention in the USA and in Southern White Vernacular English in many Southern Baptist churches in the same country. Similarly, the ethnolects, or ethnolinguistic repertoires, associated with many diasporas and transnational communities found in the UK have led to different spoken vernaculars in mosques, temples and gurdwaras throughout the country. A very basic and simple example would be the phonetic difference between a Muslim Londoner's 'brother', /bravvə/ (or /brav/), and a Muslim Yorkshireman's equivalent, /bruðə/ – a Muslim term in two variants. Although there are no historical studies in this volume, such variation inter- and intra-societally, always takes place diachronically too.

Fishman's fourth precept, regarding the co-sanctification of language varieties, also plays a part in the striving for authenticity, and its converse, the rooting out of inauthenticity within language usage. Again, the global role of English in transnational settings is a site of such striving within many faith settings with an often intergenerational tension exacerbating matters around the question of whether certain languages can be used for certain purposes, or whether they are in some way 'tainted' and inappropriate because of the role they played in imperial, colonial projects of the past and the globalising present.

In a volume like this, there is no space to go through a precept-by-precept account of Fishman's Decalogue even though most, if not all, relate directly or indirectly to any discussion of the relationship between language and religion in such a charged context as 'digital spaces'. However, precept six is worth a special mention in this introduction for its direct significance to this project. This precept states that 'all sources of sociocultural change are also sources of change in the sociolinguistic repertoire vis-à-vis religion, including religious change per se' (Fishman, 2006: 18). This is perhaps the crux of the matter as far as this volume is concerned. The sociocultural changes evinced by the arrival in the latter part of the 20th century of the internet, and electronic communications more generally, are perhaps of the same magnitude and significance as history has bestowed upon the sociocultural changes resulting from the introduction of print technology and the widespread availability of mass-produced books in the period from 1440 to 1700. The ubiquity and social domination of online practices now suggest we are witnessing another moment in history when a technology may have significant and lasting impact on all aspects of social, political, cultural and economic life, including faith practices. Moreover, as faith's presence on the internet is by far the most extensive of all topics, even outstripping sex and shopping in the number of web pages devoted to it (see Rosowsky, this volume), it is unlikely that faith will not be impacted in some way. That language is also one of the aspects of social life equally vulnerable or susceptible to this technological impact, exploration of these two together constitutes a natural rich area for such exploration.

Religion and Technology

A significant branch of the research that is taking place on new technologies and society is the attention being paid to 'online religion' or 'religion online'. In the first instance, the latter distinction between online religion and religion online has been used to distinguish between two important functions of online technologies. Firstly, 'religion online' (Helland [2000] – though see Helland [2007] for a later, more nuanced, view) is a loose term denoting digital spaces where religion has a presence associated with the internet function of information giving and has the sense of a 'first port-of-call'. These were, and still are, established as information sites, such as the Vatican's website The Holy See, where visitors can go for information, texts, explanations and the like. This is characterised as a 'one-to-many' communicative orientation. Online religion (Helland, 2000), on the other hand, makes extensive use of Web 2.0 technology and its facilitation of user-generated content and interaction – what some call 'many-to-many' communicative orientation (Campbell, 2013; Helland, 2007). Initially, this was the facility to 'chat' and discuss and synchronously and asynchronously download and upload files; however, as software designers have become more adventurous, the facility to worship and pray online, go on virtual pilgrimages and interact with teachers has become more usual and widespread across many faiths.

This development has led to an interesting symbiotic relationship between online and what is now called 'offline' religious practice, with many studies seeking to explore that relationship (Campbell, 2012) and establish patterns of similarity, convergence and divergence between the two domains. A great deal of research now exists that seeks to validate the claim that online religion is, by definition, qualitatively different to 'offline' religion and that it is not merely the unproblematic transfer of a religious practice or domain from one technological mode to another. Campbell (2013) and others explore the tension, applications and developments of these online spaces and how religious practices are seen as either simulated, authentic or transformed. An early but key article focusing on this tension is the paper by O'Leary (1996) which, drawing on Ong's (1982) thesis of the link between the introduction of print technology and the Reformation, saw fit to draw a parallel between that linkage and the recent development of electronic communication and religious practice. He drew attention not only to the way 'theories and practices of language and ritual may be profoundly altered by technological change' (O'Leary, 1996: 787), but also in what ways religious communication is able to mediate the sacred. The battles fought over the role of sacred text and image during and after the Reformation may have important lessons for understanding the manner in which the sacred is mediated by electronic and digital technology. The magic efficacy of sacred recitation of the word, mentioned by Weber (2002) and Eliade (1965), in the pre-Reformation Christian era was now replaced by the preaching of the word, with text, understood now vernacularly rather than sacramentally. This was accompanied by a parallel development in the physical environment, with church ornamentation and musical performance eradicated or severely restricted. The written texts of the Christian mission were no longer important for their sacred provenance or character but for their message, expressed in vernacular translations for the first time in a major, widespread manner. O'Leary reminds us that this change is perhaps best represented by the communion ceremony. In the Catholic tradition (i.e. pre-Reform) the words of this rite, for example, 'This is My Body', are understood performatively (Austin, 1962). Once uttered by an authorised speaker (i.e. the priest) in a suitably authorised or consecrated setting, the words become 'real' or 'true'. There is a unity of word and meaning. By contrast, within Protestantism, where the ritual was retained, the words became symbolic only, with a clear emphasis on the separation of sign and signifier and the performative character of liturgical speech acts denied altogether. This is related to the accusation that much of pre-Reformation liturgy was nothing but 'vain repetitions', memorised recitations and chanting with little or no referential meaning for those taking part, particularly within the congregation. Print technology facilitated the production of vernacular translation of sacred texts and thus, many would claim, led to different conceptualisations of religion in the Christian context, and, by extension, different understandings of language. Referential meaning now came more to the fore and language in its somatic and liturgical sense was often considered nonsensical and related by some (e.g. Calvin) to 'sorcery' (O'Leary, 1996: 789).

> We must anticipate that the propositional content and presentational form of religion in the electronic age will differ significantly as greatly from its contemporary incarnations as the teachings of Jesus differ from the dialectical theology of medieval Scholastics or as the Eucharistic ceremonies of the earliest Christians fifer from the Latin High Mass. (O'Leary, 1996: 793)

Ong writes about chirographic societies as those succeeding purely oral ones. Print technology in the 15th century led to the typographic society – McLuhan (1962) *Gutenberg Galaxy*. Electronic technology is giving us a digital society. McLuhan's proposition is that orality represented a multisensory mode of communication and interaction albeit on a physically interpersonal scale and limited to the geographically proximate. The transition to a chirographic and then typographic society reduced that multisensory mode to the unisensory mode of the visual and the sequential. Much has been claimed for this transition including the development of modernity itself. In his prescient manner, McLuhan suggested, on the evidence of film, radio and television (he was writing in the 1950s), that the visual mode of the typographic age would give way (was already giving way) to another, transformed, multisensory experience of communication and interaction. The development of electronic technologies consolidates the evidence of such a view. However, what Ong calls 'secondary orality' seems a thoroughly inadequate concept in the digital age to explain the technologically enhanced modes of communication that prevail online today. The multisensory dimension of online 'life' is a qualitatively different experience to the multisensory aspect of earlier orality. Although perhaps privileging textual practices over other practices, O'Leary reminds us of the original meaning of 'text' as something woven (from *texere*) together to make a coherent whole – words, graphics, images (still and moving), sound, colour and shape make for a superior and enhanced experience that allows for online religion, or online ritual, to have its own particular stamp – not just the usual asynchronicity and shrinkage of distance associated with online interactions and communications, but also the potential of developing online selves that differ from those of their offline counterparts. The collectivity of the offline experience, on the one hand, is mirrored by the functionality allowing multiple participants (synchronous and asynchronous), but, on the other hand, is rejected by the individual interacting virtually and, thus, physically alone. The idea of being 'alone in the crowd' particularly with the facility to 'lurk' in digital spaces becomes a strong metaphor for the experience of the online worshipper.

In her discussion of the transfer of ritual from offline to online settings, Radde-Antweiler (2006) refers to the importance of context to online religion, specifically online ritual, which is but an extension of the importance of context to offline ritual:

> ...media, the geographical, cultural, religious, political, economic, social and gender context are all aspects of context which are susceptible to change and 'a changed context...causes modifications of the ritual'. (Radde-Antweiler, 2006: 58)

She lists scripts, form, performance, aesthetics, innovation, intention reflexivity, interaction, communication, function and meaning as aspects of internal dimensions of ritual, many of which have obvious implications for modified language use. Later in this book, we will see how authors relate such contextual factors to the language of online religion and religion online, both of which feature as examples of religion in digital spaces.

As individuals can engage with online religion through online interactions, environments and intentions without ever leaving their homes, sacred cyberspace (drawing on Durkheim [1912/1976] and Eliade [1961]) is often transformed 'domestic space'. Durkheim's 'single moral community' becomes virtual, transnational and asynchronous. There is no need to be all together in the same physical space or even to understand a global faith community as made up of disparate physically separated settlements. This reminds us of Castells (1996–1998) famous definition of temporality in a networked society: 'The whole ordering of meaningful events loses its internal, chronological rhythm'. Casey (2006: 73) takes this further by commenting, 'changes in media environments produce shifts in the ways a society perceives, thinks about and believes in the world', which, in a sense, is what this book is all about – the ways in which a technology impacts upon the language(s) of religious belief and practice.

Language and the Internet

In one of the first book-length accounts of the impact of the internet on language, Crystal (2006) statement that 'if the Internet is a revolution, therefore, it is likely to be a linguistic revolution' is certainly expressing a sentiment that is shared by many. It is interesting and relevant to this volume that a list of historical precedents that Crystal cites regarding the pernicious role of technology in human affairs consists almost entirely of religious examples (e.g. the perception by the Church that printing was an invention of Satan; see Crystal [2001/2006: 2] for other examples). Granted that for much of that historical period, religion was a fundamental factor in most people's lives, but still, the point is clear. Technology, language and religion have been, and still are, inextricably linked in many ways. 'Printing enabled vernacular translations of the Bible ... adding fuel to an argument about the use of local languages in religious settings which continues to ... today' (Crystal, 2001/2006: 2). The distinctions and relationships between speech and writing are given some attention by Crystal who recognises the ambivalent nature of much online communication while highlighting significant differences between what he terms 'Netspeak' and face-to-face conversation, one example of which he cites as the conversational turn which is significantly affected by the regular asynchronous context of online communications. Examples of the internet lexicon reflected in some of the web pages featured in this volume include 'click' when referring to making some sort of physical contact in a virtual world, and the language of Facebook pages with its idiosyncratic notions of '👍 Like' and '↱ Share'.

However, Crystal's work, one of the first to seriously tackle this area, reflects his own preoccupation with language form. In the field of sociolinguistics and its focus on variety, identity and social usage, a different set of priorities have emerged. Georgakopoulou (2006) sums up the priorities for sociolinguistic research on the internet as follows:

> ...the de-localisation of interactions, the formation of communities across time and space, the performativity, transience and ephemerality of identities, the networking and interdependence of communities, the mobility and transposition of peoples, language and micro-cultures, the unprecedented flow of information and exchange among different groups that transcend the local and the national, the individuals heightened processes of reflexivity on their communication. (Georgakopoulou, 2006: 548)

Of course, much of this quotation, especially those elements to do with interdependence, mobility and flow, relates to descriptions of globalisation more generally and highlights succinctly the intimate relationship between these two factors: the relationship between a technology and a moment of historical and cultural realignment. Similarly with Androutsopoulos (2006) who, though crediting Crystal for his early comments on language and the internet, recognises that the field has moved on to debating language use rather than language form. This volume presents a range of evidence that challenges any notion of a homogeneous internet language and, like Androutsopoulos, moves away from a focus on the medium-related questions to user-related problems of language use. Furthermore, in the collection edited by Androutsopoulos (2006), it is clear that we are no longer, strictly speaking, only focusing on computer-mediated communication and that a 'digitally-mediated communication' might make more accurate sense given the explosion of mobile technologies in the past 10 years or so.

Castells' (2000: 389) notion that virtual practices work on a 'different plane of reality', of course, adds an extra dimension to the physical and spiritual planes focused on in any sociology of religion. Auer (2007: 283) reminds us that these practices are carried out through 'networks of interconnected individuals' but that despite the 'different patterns of communication and interaction' (Castells, 2000: 389) of offline communities, they are still 'real'. However, most of the communities featured in the chapters that follow have not been forged exclusively or even originally online. Indeed, most of them are both offline and online communities and groups that have migrated only partially (in most of the cases) to a digital space as communities. However, individuals within these communities have the choice, in fact, to participate exclusively online. Another traditional focus for sociolinguistics is language variation and we will see in some chapters how language variation online is often patterned not by gender or age but by doctrinal positions with certain language forms and language choices representing certain doctrinal or philosophical positions (see Rosowsky and El Naggar, this volume). Multilingualism too is reflected in the bilingual and often biscriptal approaches taken by web designers and web users as they try to navigate and negotiate their faith identities online via linguistic choices. Innovative and idiosyncratic correspondences between native and Roman graphs are a feature of many of the digital spaces appearing in this book. The vestigial orality of transcriptions rather than transliterations, where the latter presupposes a written text as central, is a characteristic of many of the websites featured in the following chapters. Finally, to complete this trawl of sociolinguistic focuses we should mention the role of English as a lingua franca of communication in so many of these transnational online worlds, a role described and analysed in more than one of the chapters that follow.

The Chapters

The chapters in this volume are diverse and far ranging in the faiths they feature and the digital platforms explored. Similarly, the linguistic observations arising in each chapter are complex and often subtle. They all, though, contribute to the theme of this book, which is the nexus of faith, language and (electronic) technology. In terms of the previous discussion, there are examples of religion online and of online religion, of offline and online convergence and divergence, of multilingualism and multiliteracies, of multisensory communication and interaction, of synchronous and asynchronous online activity, as well as some of the usual preoccupations of sociolinguistics such as the formation and negotiation of multiple identities, linguistic and performance repertoires, language variation and questions of power and discourse.

Platforms such as Facebook (and other forms of social media), YouTube, websites and Wikipedia are all featured in the following pages. Given that it is, at the time of writing, the most well-known and globally spread form of social media (though Weibo and Renren are pushing it close in the Chinese-speaking world in terms of number of users), it is no surprise that Facebook features as a platform in at least three of the chapters (Sawin, Souza and Akande) and is mentioned in most of the other chapters too.

Part 1: Faith, Language and Social Media, has two chapters both of which focus on the language practices primarily evinced by the use of social media, in each case, Facebook. Chapter 2 adopts a discursively exploratory approach and aims to show how such digital spaces allow for the affirmation of marginalised identities and negotiation and support of different Christian subjectivities. Chapter 2 draws on language planning theory to explore how transnational faith communities employ Facebook as an important online resource for developing virtual transnational networks that support language ideologies.

In his chapter, Sawin presents three quite different online/offline communities/groups (Koryo-in Central Asians, Christian academics and 'Side B' LGBT Christians), which he calls 'virtual parishes', whose online presence appears to both reinforce and resist homogenising tendencies regarding religious belief and practice. Sawin, in recognising the potential for social media platforms to be both a vehicle for minority groups seeking validation through joining with like-minded souls and a means to interact with other co-religionists whose paths they are unlikely to cross offline, reminds us of this often contradictory role of digital spaces in either confirming or extending identity repertoires. These virtual parishes reveal how many marginalised individuals and groups present within larger religious denominations find an audience for their narratives online rather than offline. The availability online of what the author calls 'dense spaces' helps to circumvent feelings of marginalisation and isolation in contrast to patterns of behaviour within larger-scaled, more dominant offline collectives. An analysis of 'speech acts of persuasion' and 'rhetorically supporting arguments', among other aspects of language practice, on Facebook, allow for a fascinating insight into how individuals and groups facing isolation find opportunities for agency and validation online. Similarly, Souza, also using Facebook as her research setting, explores the online religion and religion online dimensions within a very different faith and language community (the Brazilian Pentecostal church). The local groups she features reveal a striking fluidity and dynamism in their language practices as they negotiate the relationship between their (or others') language ideologies and their own language practices. In different language and national settings, Souza ably demonstrates the multiple layers operating within each language and faith group, many of which relate to varied multilingual practices, such as translanguaging and the continua of biliteracies. Drawing on Blommaert's (2007) notion of sociolinguistic scalarity, she explores how social media such as Facebook are used as a medium for language planning, exploring the interplay between Portuguese and other languages within the context of migrant churches. This focus on bottom-up language planning provides the reader with a complex example of how language planning and policy (LPP) is enacted in faith communities online, particularly transnational ones. Her account reveals how language planning in this bottom-up, transnational, medium is a multilingual, complex process varying through time and space. The language choices made by faith leaders reveal the various ways in which they mediate the sacred to their different audiences, linguistically, including through translanguaging practices. There is a fascinating account of how faith leaders may call in one language or a variety while their online congregations respond in their language of choice – a choice determined to a certain extent by social, cultural and national contexts. Souza reminds us how religious practice and secularity can mutually influence language choices and, in turn, be influenced by such language variation.

Part 2: Faith, Language and Transnational Online Practices, has three chapters each exploring, to varying extents, how language practice adapts itself to online contexts with transnational audiences. Chapters 3 and 4 explore the linguistic and cultural practices of the Yoruba and their transition to digital spaces. Practices here have become translocalised through the medium of the internet, allowing for significant changes to both linguistic and cultural practices in both online and offline worlds. Chapter 5, on the other hand, explores issues related to transnational conflict within the digital space of the wiki, namely Wikipedia and its entry for Yiddish, where the Yiddish Wikipedia virtual community constitutes an extension of the Yiddish-speaking diaspora more generally.

Partially drawing on Facebook, but also on one of Nigeria's longest established discussion boards, Nairaland, Akande's analysis of taboo practices among the Yoruba in West Africa is an example of the transfer of an offline, essentially oral, sacred activity to an online one, with all the issues of language form and function, appropriateness and authority that such a crucial transfer entails. For traditions 'not contained in any written law', this transfer from a context of orality to often written forms online is accompanied by implications for how a technology can impact on faith-related traditions. The focus in Akande's chapter on poetry, proverbs and other formulaic genres shows how taboo practices are transformed through the affordances of online technology linking and joining Yoruba speakers from across the globe. As an essentially oral practice, Akande explores how taboo, this important cultural and belief-based resource, is mediated via digital spaces so that it operates transnationally and globally. Among other preoccupations, she focuses on the esoteric use of language, language that is restricted to the expression of veiled mysteries and secret processes. One issue here, of course, is how the ephemerality of traditionally orally produced taboos takes on a degree of permanence, and thus a potential diminution of the esoteric, once revealed on screen.

Linking Akande's chapter to Salami's is the focus on the Yoruba people. Salami, however, presents a wider, macro-level discussion of how the Ifa religion is enacted and transformed through a range of online platforms including Facebook, mobile phones and websites. What Salami calls the 'malleability' (see Oladipo, this volume) of the electronic network widens access to sacred information. As with Akande, these oral traditions are transformed when online and, as a consequence, in this case, can lead to a potential lessening of the 'magical efficacy' of the sacred word or text, leading some to make claims for and against the authenticity of online religious practice above and beyond the basic function of information-giving (this theme links to chapters by Pandharipande and Rosowsky too). Again, there are further examples here of potential theological innovations signalled by the distinction between online religion and religion online with the novel possibility of religious advice now being communicated online as well as divination rituals being conducted using mobile technology. This chapter also provides a striking example of how the online performance of religious practice can encourage a 'personalised rather than an institutionalised narrative of faith'.

Finally in Part 2, Soldat-Jaffe's chapter returns us to some of the considerations mentioned in reference to the late Joshua Fishman at the beginning of this chapter, and neatly reminds us indirectly of his eminence in the field of Yiddish scholarship. Her fascinating account of the struggle among disparate voices for control of the Yiddish Wikipedia reminds us of the historical context for much of the complexity around language and religion. Although, as an encyclopedia, Wikipedia can be seen as facilitating the move towards 'religion online' (Helland, 2000), Yiddish Wikipedia also partakes in the world of online religion with its Web 2.0 functionality allowing for interactions and discussion between contributors and peer writers. This complex bottom-up language planning activity is at the heart of Wiki-creation more generally and leads to a tension between the competing voices seeking legitimacy for their positions vis-à-vis Yiddish. This mirrors, according to Soldat-Jaffe, the debates at large more widely in the offline diaspora community, particularly in the tension between Yiddish and Hebrew, between religion and language, and between religion, language and ethnicity.

Part 3: Faith, Language and Online Televangelism has two chapters and both explore online televangelism. The first, Chapter 7, adopts a sociolinguistic approach linked to World Englishes and explores how in digital spaces deterritorialised varieties of English can surface in novel and hitherto unexpected contexts. Chapter 8, on the other hand, takes a discourse analysis approach (like Chapter 2) to the formation and self-representation of televangelists in digital spaces as examples of how conventional distinctions between religion and entertainment are emerging online.

Omoniyi takes us into the world of digital evangelism through an exploration of online resources supporting Christian ministries with a worldwide reach and a Nigerian centre. Drawing on the fields of SLR and World Englishes, Omoniyi demonstrates how online affordances can encourage 'varieties of Nigerian English in unexpected places' (p. 135, this volume). His richly empirical study of language mobility in the 21st century provides ample evidence for a qualitative difference that is taking place where non-native varieties of English are delocalised and replace, or at least mirror, the global reach of standard varieties. Televangelism is also the focus for El Naggar's chapter as she provides a critical discourse analysis of the online performance of certain religious personalities with significant online presence and popularity. Showing how identities can be constructed via online technologies within digital spaces (particularly, in this case, YouTube), her chapter provides another good example of the multidimensional nature of digital religious practice with its characteristics of the fragmentation of religious authority and the potential for theological innovation and linguistic transformation. Adding to the analysis of the discursive strategies her subjects deploy is the recognition of the role played by the multimodal functionality of online technologies which, together, are responsible for developing a 'religio-tainment' dimension to this online practice akin to the more established genre of infotainment. The chapter, like others (Salami, Akande, Pandharipande), provides more evidence for the multisite nature of online faith practice more generally as the YouTube platform is complemented by other digital spaces such as Facebook and other social media platforms related to the central televangelists. Their performance, therefore, neatly occupies that niche between orality and (digital) literacy where the mediated communication is 'detached both spatially and temporally' from the parameters of print culture.

Part 4: Faith, Language and Online Ritual has two chapters, both of which explore the consequences of when offline ritual moves online. Language form, structure, variety and change all feature strongly in the chapters and both scrutinise theoretical frameworks of performance in their analyses.

Pandharipande presents an account of her research into online Satsang (teaching of the guru) and online puja (prayer/offering). Reminding us of the preoccupations of performance theorists such as Schechner (1988), Pandharipande problematises the notion of efficacious performance in ritual settings. Can online ritual have equal authority to ritual performed offline?

The performative language criteria – the felicity conditions – identified by Austin (1962) are either missing in online settings or else transformed. With the spiritual needs of the Hindu diaspora being somewhat hindered by physical distance and separation from the Hindu heartlands with their temples and places of pilgrimage, digital spaces provide worshippers with the opportunity to participate in ritual practices. The comparison between traditional, offline, oral Satsang and puja with their offline equivalents provides a setting for raising questions regarding form and function, authority and authenticity. Pandharipande, like Salami and Rosowsky, refers to the relationship between sign and signifier within digital spaces. The flexibility of online practices often contrasts with the fixity of offline ones. This, in itself, has implications for the signifier relationship, particularly so when the full panoply of online functionality is present – sound, image (still and moving), shape, colour – and comes into play, reminding us of McLuhan's notion of the multisensory displacing the unisensory visual.

Rosowsky meanwhile takes the lesser-known ritual of allegiance pledging to a religious teacher as his focus for explaining the impact of electronic technology on language and faith practices. His chapter reports on an analysis of website portals allowing for allegiance pledging and reveals some of the implications for language and faith when an essentially oral ritual is moved to the digital screen where, despite the affordances of multimodal functionality which can offer exclusively oral experiences, written text must nevertheless appear, embodied, visual and shaped to support the ritual. The arising uncertainties relating to original scripts, translations, transliterations, transcriptions and audio support are manifestations of the problems that oral forms face when a digital space is created for them. Coupled with the movement to a more individualised and isolated ritual experience, in contrast to its communal and collective offline counterpart, the performance of pledging allegiance becomes problematic as the efficacy of the language forms, when transferred from the authority of the teacher to the 'DIY' aspirant, become potentially diluted. Questions of authority then, and authenticity, are again very much to the fore once the oral collective ritual becomes the online digital individual act. In this respect, both authors in Part 4 offer the potential transformative effect of contextualised conventionalisation, always present, online and offline, to achieve authenticity, at least over time.

Conclusion

It is hoped that the chapters in this volume will make a modest contribution to both the field of sociology of religion and the field of sociology of language as well as being the latest offering from the developing SLR.

References

Androutsopoulos, J. (2006) Introduction: Sociolinguistics and computer-mediated communication. *Journal of Sociolinguistics* 10 (4), 419–438.

Auer, P. (2007) *Style and Social Identities: Alternative Approaches to Linguistic Heterogeneity.* Berlin: Mouton de Gruyter.

Austin, J.L. (1962) *How to do Things with Words: The William James Lectures delivered at Harvard University in 1955.* Oxford: Clarendon.

Campbell, H. (2012) Understanding the relationship between religion online and offline in a networked society. *Journal of the American Academy of Religion* 80 (1), 64–93.

Campbell, H. (2013) *Digital Religion: Understanding Religious Practice in New Media Worlds.* Abingdon: Routledge.

Casey, C. (2006) Virtual ritual, real faith: The revirtualization of religious ritual in cyberspace. *Heidelberg Journal of Religions on the Internet* 2 (1).

Castells, M. (1996–1998) *The Information Age: Economy, Society and Culture* (3 vols). Oxford: Wiley-Blackwell.

Crystal, D. (2006) *Language and the Internet.* Cambridge: Cambridge University Press.

Durkheim, D.E. (1912/1976) *The Elementary Forms of the Religious Life.* London: Allen and Unwin.

Eliade, M. (1961) *The Sacred and the Profane: The Nature of Religion.* New York: Harper Torchbooks.

Fishman, J.A. (2006) A decalogue of basic theoretical perspectives for a sociology of language and religion. In T. Omoniyi and J.A. Fishman (eds) *Explorations in the Sociology of Language and Religion* (pp. 13–25). Amsterdam: John Benjamins.

Georgakopoulou, A. (2006) Postscript: Computer-mediated communication in sociolinguistics. *Journal of Sociolinguistics* 10 (4), 548–557.

Heidelberg Journal of Religions on the Internet (2005–2015). See journals.ub.uni-heidelberg. de/index.php/religions (accessed December 2015).

Helland, C. (2000) Online-religion/religion-online and virtual communities. In J.K. Hadden and E. Cowan (eds) *Religion on the Internet: Research Prospects and Promises* (pp. 205–23). New York: JAI Press.

Helland, C. (2007) Diaspora on the electronic frontier: Developing virtual connections with sacred homelands. *Journal of Computer-Mediated Communication* 12 (3), 956–976.

Journal of Computer-Mediated Communication (1995–2015). See onlinelibrary.wiley.com/ journal/10.1111/%28ISSN%291083-6101/issues (accessed December 2015).

McLuhan, M. (1962) *The Gutenberg Galaxy: The Making of Typographic Man.* Toronto: University of Toronto Press.

O'Leary, S.D. (1996) Cyberspace as sacred space: Communicating religion on computer networks. *Journal of the American Academy of Religion* 64 (4), 781–808.

Ong, W.J. (1982) *Orality and Literacy.* London: Routledge.

Radde-Antweiler, K. (2006) Rituals online: Transferring and designing rituals. *Heidelberg Journal of Religions on the Internet* 2 (1).

Schechner, R. (1988) *Performance Theory.* New York: Routledge.

Weber, M. (1905/2002) *The Protestant Ethic and the Spirit of Capitalism* (trans. G. Wells). London: Penguin.

Part 1

Faith, Language and Social Media

2 Re-Parishing in Social Media: Identity-Based Virtual Faith Communities and Physical Parishes

Thor Sawin

The Christian ideal of parishes as 'a unity of *place* and not of *likings*, (...) bringing people of different classes and psychology together' (Lewis, 2001: 81), has seen challenges in the 21st century from mobility, neighborhood homogenization via self-sorting and niche-based church marketing. Digital social media, decoupled from geography and enabling highly specialized social organization, can both reinforce and resist these homogenizing and specializing processes. Social media allows Christians with uncommon intersectional minority identities to fellowship with those who share their specific interests (enabling new samenesses not possible within physical parishes), yet who also likely represent very different confessional or denominational backgrounds (enabling a diversity not present in physical parishes). Data from three identity-based digital Christian 'parishes' – (1) an informal Facebook network of Koryo-in Central Asians in South Korea, (2) a formal Facebook group for academics and (3) a Facebook group and blog for so-called 'Side B' lesbian, gay, bisexual and transgender (LGBT) Christians, reveals that minority Christians turn to virtual forums to fill several unmet needs. Such 'virtual parishes' created on social media platforms, while competing for engagement and affiliation with congregants' physical parishes by providing space where marginalized identities can be affirmed, also facilitate greater contentment within their physical parishes, since any unmet felt needs are met in the virtual ones. Participants engage in interpersonal encouragement, recirculate resources and authoritative voices, extend the boundaries of their physical parishes and build solidarity. The smaller scaled a particular intersectional Christian identity is, and thus the more vulnerable to exclusion or erasure in physical parish life, the more important virtual parishes are in negotiating and supporting their Christian subjectivities, and the more active its users are. While digital communities occasionally support activism to transform the ideology and practice of participants' physical parishes, they simultaneously encourage charity toward diverse physical parishes as sites of self-forgetfulness and service.

Introduction

> Thank you all for being such great examples to me of what it means to live with integrity and deep faith. Thinking about you all makes me cry from gratitude. Such beauty, love and faith lives here! Big hugs all! (Moderator of Facebook group for 'Side B' Christians)

Across the social media landscape, users' posts express heartfelt gratitude for digitally nurtured relationships. While such gratitude is widespread, for certain individuals of faith – those with marginalized or peripheral identities within their physical and local parishes – social media forums play a more significant role than expressing affection. Social media create spaces for such believers to apply their faith to very unique circumstances, circumstances not shared by many in their offline parishes, such as Whitehead's (2015) study of bereaved mothers. Digital communities in effect upscale believers' individual stories, uniting theirs with those of other similar Christians, and ultimately, as one of Whitehead's bloggers wrote, into the greater 'story God is weaving us into'.

This chapter presents three different groups of Christians who use social media to create what I call a 'virtual parish'. These virtual parishes are built around a shared identity, one that makes them a minority within the congregations they attend weekly, which I refer to as their 'physical parish'. I take a language as social action approach in documenting both the linguistic accomplishments within these virtual parishes, and how participation in these virtual parishes reframes participation in their physical parishes. Tensions between promoting diversity and seeking similarity, a key issue in Christian ecclesiology, play out in participants' discursive construction of their unique positioning as Christians within complex minority groups.

Theoretical Background

Before introducing the particular populations I am analyzing, it is worth reviewing some of the larger issues around minority identities, religious communities and online performances of identity and faith.

Christian identity

Since 'Christian' is a contestable term, it is worth defining how that term is applied in this study. Canagarajah (2004) classifies identities into those based on *biography* – the product of an individual's history (gender, race, nationality), on *role* – a temporary social position (teacher, student, friend) and on *subjectivity* – a positioning vis-à-vis stable discourses which embody the ideologies of a society. In this chapter, I do not use 'Christian identity' in the *biographical* sense – someone Christian by virtue of birth, heritage, social or relational milieu. I treat the term as a *subjectivity*, whereby individual agents actively align with and ascribe to themselves the discourse of Christians. Subjectivity-based identities, such as 'lazy immigrant' or 'Evangelical Christian' can be ascribed by another actor, but more important is whether and how an individual agentively adopts that subjectivity, and aligns himself/herself with its attendant, socially circulating discourses. Christianity as a subjectivity is largely accomplished through language behavior (Bucholtz & Hall, 2004). The authentication of a Christian identity, like any other identity, depends also on the uptake of others who claim that identity and whether they vouch for or delegitimate one's claims to being Christian. The Christians in this chapter are ones who attempt to organize their lives around enacting 'Christian' in word, thought, and deed, and who are ratified as Christians by authorities within their parish.

Religious parishes and diversity

I intentionally use 'parish' (rather than 'congregation' or 'local church') to refer to stable communities that gather physically and regularly under the same local leadership. The term 'parish' combines notions of temporality (shared times), locality (shared spaces) and affiliation (shared experiences). In the early Christian centuries, each locality had one common gathering for all its Christians, making each local parish as diverse as the local Christian population. Paul's instructions to early church leaders (Colossians 3:11, 1 Peter 2:5) uphold this ideal. In the medieval and modern eras, as Christians became more numerous in cities, parishes became ever more local and, by mirroring neighborhood segregation, less diverse. Yet, in the 20th century, C.S. Lewis (2001: 81) still famously defended Paul's parish ideal against the growing tendency to market churches to specific demographics, valuing diverse local parishes as 'a unity of place and not of likings, (...) bringing people of different classes and psychology together'. Several studies have documented how even in the 21st century, many American congregations play this role of bringing ethnically (Marti, 2009), racially (Edwards *et al.*, 2013) and politically (Bean, 2014) diverse Christians into symbolic and physical interaction with each other. The adjectival form *parochial*, today meaning 'of limited perspective or diversity', ironically signifies the very attitude that locality based parishes were meant to guard against.

Yet, since the mid-20th century, mobility at all scales has increased, while neighborhoods have homogenized via self-selection. This mobility unhinges the demographics of a particular neighborhood from the demographics of a church located within that neighborhood, as people travel outside their own locality to attend church. According to Brunn *et al.* (2015), neoliberal discourses have also proliferated, which envision a marketplace of demographically niched churches to be consumed by 'church shoppers' whose attention and allegiance needs competing for. Whereas early Christian churches and the Anglican parishes of C.S. Lewis' day were highly local in scale, hosting congregants of diverse classes, professions and persuasions; in the 21st century, Christians often travel great distances in order to worship with a congregation niched to their taste – the 'unity of likings' that Lewis opposed. Ironically then, the decline of geographically organized *parishes* has led to greater *parochialism*, by decreasing *intra*-parish and increasing *inter*-parish differences. The increased role of social media in the lives of 21st-century Christians plays out against this background tension. Is the Christian faith practiced more effectively in highly diverse communities, or in communities made up of 'people just like me'? While the internet seems to be the ultimate vehicle whereby Christians of very diverse backgrounds and experiences might encounter each other, by eliminating geographic distance as a constraint, the internet conversely enables believers to spend more time engaging with only those already most 'united in liking', mirroring their own experience.

Minority identity theory

Minority identities are not merely multifaceted, but nested and intersectional. *Intersectionality* presupposes that being a minority-within-a-minority (i.e. a black female) is qualitatively different than the sum of two separate experiences of being a minority (being black plus being female). Such nesting invokes the sociolinguistic concept of scale (Blommaert *et al.*, 2015). Scale, 'semiotized space and time' (Blommaert *et al.*, 2015: 3), refers to the boundaries within which a given sign has certain denotational or indexical meanings. Scale can refer to bounded and nested expanses of discourse, of time, or space across the surface of the earth, and the 'scope of comprehensibility' (Blommaert *et al.*, 2015: 5) for a sign often varies in all three dimensions at once. The larger the scale of a sign-meaning mapping, the more unmarked, or 'presupposeable' (Blommaert *et al.*, 2015: 5), its deployment is.

I focus here on geographical scale, the scale of an identity as a function of both how widely and how densely individuals claiming that identity are distributed across a surface. For example, PhD holders are more *widely distributed* across the United States than are Karen refugees, and each group can be either *densely concentrated*, such as in university towns or designated resettlement centers, or *sparsely concentrated*, perhaps in a rural town. Individuals feel vulnerable within an identity (or 'thin-skinned' in linguistic encounters [Meyerhoff & Niedlzielski, 1994]) to the extent that the identity is smaller scaled – not widely distributed and/or sparsely concentrated – and thus overshadowed by larger-scaled, more dominant and powerful identities. Worldwide, Christianity is often experienced as a minority faith. For Christians who also identify with a racial, ethnic or sexual minority, their faith is experienced as an intersectional 'minority within a minority'. Indonesian-expat Christians, for example, are in a double bind; practicing Christians are a minority in the expat Indonesian community, while Indonesians are also a minority within their host countries' Christian community. Even in the non-rural West, Christianity can feel like a smaller-scaled subjectivity dominated by an increasingly secular matrix culture. I analyze virtual parishes as a space for 'upscaling' particular sign-meaning correspondences. Although reliant on the computer programming that underlies the capability to create such spaces, identity work is done primarily via the linguistic action of members. As those dispersed individuals collect to create a screen space rich with instances of that mapping, a sign with a scattered and sparse scale can become concentrated and dense, which gives it the appearance of being frequent, common and possibly even dominant within that space.

Creating spaces with a higher density of occurrence then partially alleviates vulnerability. This impulse, which prompted the creation of Christian parish meetings in the first place, is observable in geographic space as ethnic neighborhoods (Vervoort et al., 2012), 'gayborhoods' (Ghaziani, 2015) or religious enclaves (Cimino, 2011), and in the virtual realm takes the form of digital forums –'dense spaces' within which one particular matrix of intersecting identities can become the dominant one. The virtual space may be especially important if that axiomatic identity (around which the space is constructed) is rare, marginalized in the physical world. The following Facebook groups are dense spaces for individuals with particular intersectional minority identities:

(1) Christian Biker Unity South Carolina
(2) Deaf Christian Association of Nigeria
(3) Latin Jews DC

For these individuals, social media spaces permit normally rare intersectional identities, with limited physical space for expression in the offline world, to emerge as a coherent identity in its own right. Insufficient local demand for a physical Latino Jewish synagogue in Washington DC or for a deaf Christian camp in Nigeria is no obstacle to the creation of virtual communities. Through digital spaces, not only is the existence of such individuals called to the world's attention, a heretofore invisible identity is also called 'into being' – granted a social life and a visible form.

Characteristics of online religious practice

Zhao et al. (2008: 1817) noted that the disembodied and relatively anonymous nature of social media facilitates 'a new mode of identity production'. As individuals compose, post and respond to each other, they create and reinforce a model for deploying sign-meaning mappings at several levels – from terminology choice on the micro-level to macro-level patterns to what social ends language should be deployed – within that virtual parish, and by extension among individuals with a given intersectional identity. The linguistic social practice, the ends to which language is put, form the primary unit of language analysis in this study.

Campbell (2005) found that in the mid-2000s digital religious practice took two main forms – forming relationships with co-religionists, and seeking specific spiritual information online. Both these relationship-building and information-seeking functions are affected by Suler's (2004) 'disinhibitory effect'. Internet disembodiment can create *toxic disinhibition* and behaviors such as flaming, trolling and caricaturing. Yet, the relative anonymity and lack of felt social consequences which attend online interaction can also have positive effects, known as *facilitative disinhibition*. Individuals are free to seek information on taboo topics, explore marginal positions, or 'come out' in order to gain solidarity and support. For committed Christians, strong moral constraints on stance-taking minimize the likelihood of toxic disinhibition, leading to the prevalence of facilitative disinhibition. Such facilitative disinhibition, coupled with social media's affordance of transcending restricted distribution or sparse concentration, makes social media especially well-suited for receiving encouragement via online language. For example, Taylor *et al.* (2014) documented how queer Christian youth seeking support from other experienced or authoritative voices come together online to navigate the fraught space between their religious and sexual identities. Another example is Whitehead's (2015) study of mothers of deceased infants, who communed online with other individuals undergoing that comparatively rare circumstance. The internet both enabled these women to locate others and freed them to discuss a topic still largely taboo and discomforting to other Christians in their physical parish. These two studies also illustrate Campbell's (2012) list of five consistent traits for religious expression in social media: networked community (high degree of interactivity); storied identities (identity creation via experiential narratives); shifting authority (nominating alternative, non-vested authoritative voices); convergent practice (around new norms of practice created by authority figures); and multisite reality (interplay of online and offline interaction). Participation in such virtual communities allows personal religious practice to transcend physical parishes, and includes believers who exhibit previously unencountered forms of diversity and similarity.

Methods and Data

The data for this chapter comes from three different groups of Christians, each forming a sort of virtual parish. Each parish is one in which I was a legitimate, peripheral member, in a capacity explained within each of the following sections. For data collected from established forums,

I approached the moderators of the forums for permission to gather data and answered any of their concerns about my purposes and privacy. Once the moderators' approval was obtained, I observed each group for several months in order to ascertain the general patterns of how core/active and peripheral/less-active members contributed to the spaces. Once habitual practice was identified, I next selected a period of time within which to read and analyze the posted language. User language was pasted into a separate document for analysis, removing all identifying information from individual examples, and replacing any names with pseudonyms. I felt this especially important for the Side B Christians, whose identity is arguable the most vulnerable. For each individual post cited in this chapter, or which played a formational role in the analysis, I approached the individual poster by describing the nature of this study and asking whether I could use his/her quote (including the exact wording [with all pseudonym-use]) as it appears in this chapter. I did this even for the blog posts that are positioned as 'public' – accessible to any reader with the URL. None of the users whom I contacted about using their data refused the request, and all were eager to participate.

Evangelical Koryo-saram

The first dataset is an informal network of Evangelical Christian Koryo-saram originally from Central Asia, but living in South Korea at the time the data was collected in spring 2011. I was an informal faculty advisor for this group from 2006 to 2008, and nurtured an ongoing friendship with them after my formal responsibilities ended. Koryo-saram identity is complex – ethnically Korean Russophones whose ancestors were deported by Stalin to communities scattered around Central Asia (Kim, 2003). Many in the post-Soviet generation, like those in this study, have left their home communities to pursue opportunities in European Russia, South Korea, China and the United States.

For this reason, Facebook in 2011 was the primary 'place' where expatriate Koryo-saram moving throughout Russia, Central Asia, China and Korea performed their solidarity, and commiserated about their shared experience of being a perpetual 'other'. This diverse repertoire of Russian-Korean-English translanguaging along with their distinct cultural background, make the Koryo-saram stand out whether attending Russian-mediated, Korean-mediated or English-mediated parishes.

These individuals became Christians in their youth through contact with South Korean missionaries, and subsequently attended Christian universities. The data consists of public Facebook postings (visible to their entire friend network) and comments to those posts. Their Facebook interactions represent the digital persistence and merger of social groups established in those physical parishes in their formative years at university. Unlike the other datasets, these posts do not come from a defined forum, but from public Facebook interactions, and thus are visible both to non-Koryo-saram and to non-Christians. While *82Avenue* is a dedicated Facebook group for Russophones in Korea, no equivalents exist specifically for Koryo-saram or Russophone Christians.

Faith in the Academy

The second dataset consists of posts from 2013 to 2015 by members of a closed Facebook group called 'Faith in the Academy', coterminous with a registered student organization at a large American public R1 university. The posters are graduate students and faculty, along with individuals from the city interested in the intersection of Christian faith and academic scholarship. The group consists of Christian academics, an intersection underrepresented within its two larger superset groups – Evangelical Christians and mainstream academia. The mission statement reads:

> We exist to encourage one another to faithfully pursue our calling as disciples of Christ in our academic and professional roles, for the betterment of ourselves, the benefit of our community and the display of God's glory.

Faith in the Academy data differs from the Koryo-saram data in two key ways: it comes from a closed and moderated Facebook group and it is generated by individuals who live in the same locality. Although the group does sponsor regular physical meetings, which I attended from 2011 to 2013, many members of the Facebook group (which the author still belonged to as an alumnus) cannot attend and most of the key content sharing and interaction exists only in digital form. Members of Faith in the Academy are simultaneously members of several different physical parishes across the city. Thus, the group becomes an avenue for physical unity among members of various diverse parishes, as well as creating connections between members of the same parish. Crucially, the Facebook group is not merely the established digital footprint of a primarily physical community, but also a virtual community producing a substantial physical footprint of sponsored events and discussions.

Side B Christians

The third dataset involves Christians who self-identify as LGBT or same-sex attracted, yet whose interpretation of Christian teaching prohibits same-sex sexual relationships. Members are thus either celibate or in mixed-orientation marriages. Such individuals are referred to among LGBT Christians as 'Side B', with 'Side A' denoting those who believe Christianity permits same-sex relationships. For this group then, a biographical identity – being LGBT – is subsumed by a subjectivity-based identity – being a celibate Christian. Side B Christians represent a small subset of self-identifying LGBT Christians, which is itself a minority within both the Christian and LGBT communities, each a superset seen as a minority with society as a whole. On one end of the Side B spectrum are those with primarily LGBT friends, a looser connection to physical parishes and few Christian friends with whom to discuss LGBT issues from a Christian perspective. On the other end are those with primarily conservative Christian friends, a stronger connection to physical parishes and fewer LGBT friends with whom to discuss Christian issues from their LGBT perspective. The fact that Side B is a highly specific intersection of identities – Christian, LGBT, celibate or heterosexually partnered – paradoxically leads to great diversity within Side B forums in other dimensions. Contributors to both of these digital forums are scattered around the world, and represent a rich diversity of Christian confessions (i.e. Protestant, Orthodox, Catholics) and denominations (i.e. Baptist, Pentecostal, Reformed). Side B forums expose members often for the first time to Christians from very different theological or cultural backgrounds; shared sexual orientation leads to marked ecumenism. Members also diversely self-identify as gay, lesbian, bisexual, transgendered and/or same-sex attracted. The novelty of this diversity of sexual orientation is frequently noted by participants, who have less exposure to all these sub-communities in their offline lives.

Data for this group, collected in the fall of 2015, comes from two different forums, a closed Facebook group, which I was invited to join and vouched for by several well-positioned members, and a publicly accessible blog. The Facebook group is a closed group, wherein ratified Side B members interactively comment on each other's short posts, often 'friending' each other across social media. Contributors' online personae are thus more linked to their 'real life' identities than the commenters to the Side B blog. This blog is curated, with essay-length contributions solicited from a vetted set of bloggers, and open to anonymous commentary. The ostensible audience for the blog is other Side B Christians, but members of 'neighboring' groups such as Side A or non-LGBT Christians also comment. The overarching purpose of the blog is to encourage readers to find relational fulfillment in Christian friendships, equipping them to do so and troubleshooting their attempts. Posts tend to be speech acts of persuasion, rhetorically supporting arguments through disclosing personal narratives.

Summary of datasets

Table 2.1 summarizes the characteristics of the digital parishes for the three groups of Christians in this chapter.

Table 2.1 Characteristics of the three groups used as data in this study

	Koryo-in	Faith in the Academy	Side B	
Key shared identity	Geographic origin+ethnicity	Academics in a state university	Celibate or heterosexually partnered LGBT	
Group is a minority within these groups	All Koryo-saram All Russophones in Korea All diaspora Koreans All Central Asian citizens	All Christians All academics	All LGBT Christians All LGBT All Christians	
Text type	Open Facebook postings	Closed Facebook group + email listserv	Curated blog	Closed Facebook group
Purpose of group	Casual social inter-actions, no stated purpose	Encouragement and idea-sharing	Philosophical, theological and practical short essays for encouragement and persuasion	
Relation to physical parishes	Belong to various Korean and Russian-medium evangelical parishes in Korea	Belong to sev-eral evangeli-cal parishes in the same city	Range of parishes around the world and across the Christian confessional spec-trum. May be wary/skeptical of physical parishes	

Findings

Such diverse groups of Christians might be expected to share little in terms of goals and practices in their digital faith expressions. This study explores which linguistic acts minority Christians perform when exposed in virtual spaces to others 'just like me', and how such spaces are related to their experience in physical parishes that may not fulfill their specialized needs. Campbell (2005) identified information seeking and relationship building as the two metafunctions of the religious internet. She later noted such practice as further characterized by networked community, storied identities, shifting authority, convergent practice and a multisite reality (Campbell, 2012). My datasets of minority Christians reveal an overlapping, yet significantly different set of primary functions:

(1) Encouraging other believers.
(2) Ventriloquating (Bakhtin, 1981: 293) authoritative voices.
(3) Extending physical parishes.

(4) Producing solidarity via stance-taking toward a non-present group of 'others'.

This section elaborates on and illustrates each behavior, discussing why the different groups might be engaging in those practices.

Encouraging others

The stated purpose of the Koryo-saram and Side B forums is to encourage believers with a marginalized or vulnerable identity. That encouragement and affirmations should make up a large part of the discourse in these digital forums is thus unsurprising, and in line with Campbell's observation that digital spirituality is substantially about relationship building. Encouragement may come to the fore because Koryo-sarams, academics and especially Side B LGBT believers are vulnerable to feeling underrepresented or misunderstood in their physical parishes.

The Koryo-saram have access to a much wider range of spiritual content than the mono- or bilinguals in their physical parishes and affirmation consisted primarily of general Christian encouragement, but drawing on their uniquely broad repertoire of resources, situated at the intersection of Uzbek, Korean, Russian and Western identities. Encouragement for them often takes the form of links to videos of worship songs in Korean, Russian and English, and repostings of Bible verses and inspirational messages in all three languages. By deploying this content on Facebook, they simultaneously discharge their Christian duty to encourage and build solidarity through commentary and likes, and they also call their unique Koryo-saram trilingual identity into relevance; only other Russian–Korean–English trilinguals would be maximally encouraged by the content they post. Encouragement can also reference their shared origin, as in Pavel's following post:

> Что я понял живя в Узбекистане? Что Иисус живет в Узбекистане!
> *'What did I learn living in Uzbekistan? That Jesus is alive in Uzbekistan'!*

As Christians of many kinds face increased persecution in Central Asia, this may be a note of encouragement to those worried about the challenges their physical churches are facing. In the following posting, Dasha and Roza create solidarity while demonstrating that they find daily spiritual sustenance in a non-native tongue from an English-language devotional *Our Daily Bread* (ODB).

> **Dasha** (post): He shall give His angels charge over you, to keep you in all your ways – Psalm 91:11

Roza (reply): ODB?:) I was also meditating on this verse:) very encouraging!

Dasha (reply): ODB ☺ It encouraged me too.

Dasha and Roza are typical of the Koryo-saram in that their private faith practice draws heavily on both English and Russian resources, only occasionally do they reference Christian materials in Korean, the main language of their spiritual milieu. In Figure 2.1, Pavel posts an image, liked by 11 other Koryo-saram, taken from the Russian social network *vKontakte*, specifically a forum entitled *onlayn-tserkov* ('online church').

'Все ждут конца СВЕТА' ambiguously means both 'everyone is waiting for the end of the WORLD' and 'everyone is waiting for the end of the LIGHT'. This wordplay sets up the glowing second half of the quote 'Я жду конца ТЬМЫ' (*'I'm waiting for the end of the DARKNESS'*), communicating his hopeful stance about the ultimate triumph of light. Koryo-saram like Pavel, an Uzbek citizen in Korea, circulate and engage in creative ways with online Russophone Christian resources, filling a gap left by their primarily Anglophone and Koreaphone physical Christian community.

Because Side B identity is nested within so many superset layers of minority identity – Christian, LGBT and LGBT Christian – encouragement and affirmation are particularly dominant in online practice. Side B Christians generally presuppose, as Bozard and Sanders (2011) claimed, that submission to and integration within a local church is of great benefit. Yet, a preponderance of previous work claims that for LGBT people, Christian belief is harmful (Gross, 2008; Kubicek *et al.*, 2009), superficial (Gross

Figure 2.1 Pavel posts an encouraging image containing Russian wordplay

& Yip, 2010) or requires dismissing church teaching (Hall, 2015). The Side B community is thus vulnerable to challenges on several fronts: from conservative Christians who disagree that Christians should self-identify as LGBT at all; from the large portion of the LGBT community suspicious of religious belief in general; and from Side A LGBT Christians who claim that Side B insistence on celibacy is equally as harmful as 20th-century Christian rhetoric of orientation change. Vulnerability is exacerbated by the fact that few have close *physical* friendships with other Side B Christians in their physical parishes, something frequently commented on when discussing *virtual* ones:

> One thing that came up [*in a Skype session with another forum member*] was the kinship of LGBT Christians and how it is a really tremendous blessing that I cannot thank God enough for... (Side B forum user)

Many different challenges lead Side B Christians online for support. Participants' posts confess depression and isolation, report conflicts with others, express frustration with widely circulating rhetoric and document everyday awkwardnesses of being Side B; all are met with enthusiastic encouragement and offers of prayer. More positively, participants report opportunities to come out to close friends and family, or to assist their physical parishes in becoming more helpful to the same-sex attracted. Participants also frequently post 'inside' jokes and memes originating in the wider LGBT community; such humor lightens the mood of otherwise heavy conversations. LGBT-themed humor may be especially prevalent since opportunities to joke in such ways about their particular trials may be rare among the non-LGBT Christians in their physical parishes.

For the Faith in the Academy group, although encouragement is central to their mission statement ('we exist to encourage one another to faithfully pursue our calling as disciples of Christ in our academic and professional roles'), encouragement takes the form of overt posts less frequently than among Koryo-saram or Side B'ers. Perhaps since the Faith in the Academy group has regular physical meetings, unlike the other two groups, encouragement is more likely to occur face to face. Encouragement online took the form of mobilizing essays or articles published by more authoritative voices, a practice discussed in the next subsection.

In summary, the frequent acts of encouragement indicate that the virtual parish meets a need that physical parishes cannot. Someone with shared personal experience may be more effective at offering encouragement that 'land' than well-meaning believers of outsider groups. While diverse other Christians from their physical parishes almost certainly attempt encouragement, a special role remains for believers who personally identify with the struggles experienced at precisely the same intersection of minority identities.

Ventriloquating authoritative voices

Campbell (2012) found that certain participants constructed an authoritative role for themselves within online spiritual communities, creating centers of authority which rivaled those found in the traditional church structure. In this data, however, the most authoritative voices tended not to be those of the individual participants posting in their own words. Rather, participants frequently reposted, excerpted or cited voices from Christian theology, social commentary or literature via links to websites, blogs and videos. Bakhtin (1981) referred to this as ventriloquation – when a person communicates an idea that he/she endorses and agrees with, but uses the cited words of another (often a more authoritative voice) in order to deploy and add more weight to those ideas. This takes the form of either posting a link to an intact discourse available elsewhere online, or posting meaningful excerpts of such discourses which accomplish a more specific social action (such as challenging or encouraging). Ventriloquating a famous voice is a fraught discursive move, as one could seem to be stealing that voice's authority, boasting about how well-read one is, or other acts deemed prideful. For this reason, posters often preempted such interpretations in an introduction using their own voice to frame the quote, thereby adopting a subservient posture toward the ventriloquated voices:

'very moving testimony by...'
'anyone else feel this a lot'?
'hope its an encouragement'
'this is a wonderful sermon'
'very helpful article for me this past week'

Ventriloquating more widely published voices accomplishes several things simultaneously. First, the media linked often serve the encouragement function described earlier. Secondly, discovering that well-known writers have written eloquently on topics of interest to that particular minority group may lessen feelings of invisibility or vulnerability among that minority. Thirdly, the 'information-seeking' metafunction identified by Campbell (2005) is facilitated, as participants encounter curated links to sermons, books and articles of interest. This was particularly true for the Faith in the Academy group, where a veritable library of links of interest to Christian academics are posted and endorsed. Indeed, linking insightful media seems to be the primary online method for the encouragement spoken of in their mission statement. By linking essays, articles or excerpts from books by well-known Christian academics like C.S. Lewis, Thomas Marsden or Mark Noll, participants introduce more authoritative, less marginal voices into their virtual community, bolstering their right to exist and influencing both the larger Christian and academic worlds.

This act of animating the authoritative voices of others by linking media also creates new centers of authority within the community. For example, one self-described avid reader has become an appreciated contributor by frequently posting inspirational excerpts from Christian writers on themes of longing, self-denial and loneliness in the Side B Facebook group. Another Side B poster, well known for his own blog, lent authority to others in the community by attempting to publicize their personal blogs:

> Side B bloggers: A number of people in the group are bloggers. Please list blog names below!

Information on 17 blogs was then added below by participants, including the most prominent Side B blog, the additional data source in this study. The vetted authors from that prominent blog have become the 'authoritative voices' whose writings are frequently ventriloquated and discussed in the Facebook group. When one of those authors – most of whom also belong to the Facebook group – posts, his/her posts carry weight by virtue of his/her public position in the larger church's discussion of homosexuality, and his/her resulting personal connections with leading Christian figures. The individuals whose blogs were listed in the above post derived authority both from the reputation of the man who initiated this effort and from being listed alongside these highly respected authors. This demonstration of the quantity and quality of Side B blogs both serves the information-seeking function and solidifies the sense that there is a considerable body of published Side B thought.

In summary, this ventriloquation of authoritative voices does not merely inform, resource and encourage believers with their content. The act of animating more authoritative Christians can make marginal identity combinations within a given physical parish feel more solid and less invisible. That well-known Christians 'see' these identities and speak to their circumstances helps demarginalize them by conferring an identity legitimated within the larger Christian community.

Extending physical parishes

Campbell (2012) noted that a 'multisite reality' was a hallmark of the spiritual internet. Indeed, a clear distinction between physical and virtual parishes is not always sustainable; interactions in the virtual world connect and carryover into the physical, resulting in a multisited Christian experience. Physical parishes often host their own Facebook page or website, and virtual forums such as those studied here can lead to new physical communities.

Boundaries between physical and virtual parishes get blurred in two key ways. First, virtual parishes can lend an 'afterlife' to physical parishes for former members removed by time and distance. This was especially true for the Koryo-saram. The social connections among Evangelical Koryo-saram in Korea were forged in physical parishes in their Central Asian hometowns or universities. For example, Ivan, living in Korea, stays connected to Andrei, a friend still attending his old parish in Central Asia, where preparations for a congregational anniversary celebration are underway:

Andrei: Хай мэн!   Как поживаешь?
        khai men! (Eng. 'hi man') how's it going?
Ivan: здорово (...) У тебя как? как у вас там с ювелиркой дела?
        great (...) You? How are things going there with the anniversary celebration?
Andrei: u nas vse ok. Uvelirka tiho tiho idet. Rabotaem poka)
        here things are all good. The celebration is coming along. Still working on it)

Ivan is able to remain a part of his physical parish through virtual links across time and distance.

Also, among Faith in the Academy members, graduates of the university occasionally visit the Facebook group, expressing thankfulness for past experience, and to avail themselves of the resources posted. In this way, the Faith in the Academy group also creates an 'afterlife' for distant/former members.

Secondly, virtual parishes are used to negotiate physical meetings, meetings parallel to members' regular church services in their physical parishes. Such meetings are sometimes called 'parachurch' meetings, supplementing yet subordinate to the regular parish life of Christians (Hammett, 2000), and centered on a specific issue rather than filling all the functions of parish life. The majority of posts to the Faith in the Academy group, who unlike the other groups resided in the same city, consisted of invitations to and arrangements for physical meetings: special lectures, roundtable discussions and social happy hours. These physical meetings crosscut membership in various physical parishes across the city, creating new opportunities for fellowship and group affiliation. Unlike Koryo-saram and Side B'ers, Faith in the Academy members turned to the website less for interpersonal interaction, and more to disseminate vetted resources while advertising and sharing insights from these parachurch gatherings.

For the Side B group, occasional physical meetings are used to sustain relationships forged in the virtual forum. One form this took was the decision to organize an annual retreat in the United States. Although the participants are geographically quite scattered, many travel great distances – even internationally – to attend the retreat, confirming the significance that this virtual parish has in their lives. Besides the retreat, participants who discover from the profile information that they live in geographical proximity to others also occasionally meet up, a picture from which was described in one post:

> Spending Sunday with the gay guys !! Worshipped together at the local [name of] Church where we spend some really important time praying for each other. This was followed by lunch and then frolicking around on the (very windy) peninsula.

In summary, virtual parishes cannot exist without affecting relationships in the physical world. Whether the virtual parish is extending connections originally forged in physical parishes, or facilitating brand new parachurch meetings, virtual parishes create new 'belongings', allegiances which allow believers to transcend their local physical parishes. The scope of users' physical fellowship is thus expanded to include attendees of very different parishes, in the same city and around the world.

Solidarity as against the 'other'

In sociolinguistic theory, solidarity is created as users take similar stances (i.e. praising, criticizing, mocking) toward the same set of stance objects (i.e. artifacts, people, linguistic performances) (DuBois, 2004). Collections of stances can coalesce into stable practice, as new or peripheral members of a community get socialized by more established members, learning the kinds of stances and opinions that members of a group are supposed to express. This well-documented process of socialization, what Campbell (2012) labeled 'convergent practice', takes on special characteristics in this study's data. For the Koryo-saram, Faith in the Academy and Side B, all of whom are minority groups nested within other larger minorities, group solidarity often emerges via stance-taking toward another non-present group, which serves as an identity foil. This group is often one of the larger groups within whom these smaller minorities are nested, and against whom their identity ostensibly needs defending. The groups that act as a foil are laid out in Table 2.2.

Table 2.2 Groups which act as a foil for the three data groups

Group	Group acting as a foil (frequent negative stances)
Koryo-saram	Koreans (as a foreign, bewildering or constraining culture)
Faith in the Academy	The academy (as hostile towards faith) The church (as skeptical of academia)
Side B	Conservative evangelicals (as ham-handed regarding LGBT issues)

Koryo-saram identity, as noted in Table 2.1, is nested within several larger identity groups – Central Asians, Russians and Koreans. Although Central Asian governments' regulations on expatriates are frustrating, Central Asia is rarely a stance object in Koryo-saram Facebook posts. Russians as a group are also noticeably absent from their discourse, perhaps because, despite their Russian mother tongue and Russian first names, few have had much contact with Russia or ethnic Russians. Koreans, of all their dominating identity groups, appear most frequently. In Korea, Koryo-saram frequently encounter small instances of culture shock, differing norms for workers, behavior and gender roles which can bemuse and bewilder Koryo-saram raised in a non-Confucian (post-)Soviet culture. Comments about Korea(ns) are almost always made in Russian, posting 'behind the back' of South Korean Facebook friends bilingual in Korean and English. However, there are no instances in the dataset of a negative stance toward the Korean church or Korean Christian practice. Since most of the Koryo-saram's Christian experiences have involved Korean-trained missionaries and pastors, it is perhaps unsurprising that they are more charitable toward the Korean church than other aspects of Korean society.

For the Faith and Scholarship society, the fact that the academy is hostile to Christian references or impulses seems to be a given. The forum moderator linked a video lecture by Mark Noll entitled 'Christianity: Expanding Worldwide yet Struggling in Universities. Why?' and an article by David Williams called 'Why You Must Be Dying to be a Christian Scholar' – both lamenting the hostility that Christians face in academic settings, which are described as assuming faith to be unscientific and anachronistic. Moves which position academia as a hostile and difficult environment, where Christians need constant encouragement to thrive, go little challenged or commented on, such as the following post occasioned by the Day of Prayer:

> National day of prayer. Prayed for you all this morning. please know you encourage me in you [sic] boldness of faith in a secular academic world. May I be so bold and honor God as you do.

Yet, besides the discourse of academia as hostile to faith, certain posts position the church as being reciprocally hostile to academics. One participant quoted from Kinnaman (2011):

> Millions of Christ-following teens and young adults are interested in serving in mainstream professions .... Yet most receive little guidance from their church communities for how to connect these vocational dreams deeply with their faith in Christ. (Kinnaman, 2011: 29)

Another quoted from Berlinerblau (2005: 73), a non-religious author writing about the church's skepticism of Christian academics:

> Unappreciated by their secular colleagues in the Academy .... they receive the less enthusiastic handshake, the guarded, tensile hug that has been perennially reserved for those afflicted by subversive knowledge. (Berlinerblau, 2005)

In contrast to moves which position academia as a hostile overarching group, attempts to position the church as a hostile overarching group get contested. One member personally responded to the above quote:

> I couldn't possibly disagree with this any more than I do. To suggest any level of persecution from within or without the church, against academics, is demeaning to those before us who actually have been persecuted. Plus, it goes against my regular experience as a Christian academic in my own local church.

In the Faith and Academy group, as for the Koryo-sarams, negative stances toward the church or other Christians seem to be largely avoided.

The situation for the Side B group is more complex. Participants frequently allude to feeling pressed between two competing and hostile groups – a larger LGBT community skeptical of Side B adherence to orthodox teaching on sexuality, and a larger church community skeptical of Side B self-identification as LGBT. If the pattern found for Koryo-saram and Faith in the Academy held, one would expect the Side B'ers to be more charitable toward the church, and take more negative stances toward the secular LGBT community. Interestingly, this is rarely the case; almost all negative stances seem to be taken toward the conservative, evangelical wing of the Christian church, especially against non-LGBT evangelical writers who attempt to write about LGBT issues and the consequences of increasing societal acceptance of homosexuality. One example, wherein the rhetorical failings of a noted Christian author get repackaged as moral failings of and by the Side B poster, is

> Lord forgive me for my anger. Even though some of his writing have [sic] good parts, so often when I read what [name of noted Christian author] [sic] I just can't help but to want to punch him so hard in the face. I strive for pacifism but fail so hard at it.

There are two possible explanations why well-known conservative Christians are seen as a greater foil than the LGBT community. First, participants in the Side B forum represent a range of personal experiences. Some individuals report being very strongly connected, socially and spiritually, in their physical parishes. Such individuals usually have straight church friends aware of and encouraging about the issues arising from sexual identity. This strong support in the physical parish tends to relegate the virtual parish of the Side B forum to merely an occasional source of extra resources and support, and such individuals tend to be 'quieter' on the forum. Other participants report really struggling with or feeling alienated from their local church and/or from Christianity at large. For such individuals, the virtual forum becomes perhaps the sole site of support, and sole safe place in which to vent frustrations – raw frustrations often backpedalled or nuanced in later commentary.

Secondly, the particular Side B forums studied here are not the only ones on social media. Other groups exist that are aimed at those ambivalent about being identified with the LGBT community, and more freely identifying as conservative Evangelical Christians. The particular Side B communities studied here draw more heavily from mainline or traditional Christian traditions, which are more skeptical of conservative evangelical rhetoric in general. However, the act of taking strong negative stances toward a non-present other, while an effective way to build in-group solidarity, is a problematic move for Christians, as the following post acknowledges. In the first part, the poster acknowledges the tension of being caught between critics from two dominating groups:

> I wanted to voice something that's concerned me a little about the way we sometimes react to our critics – Christian or queer. They're often extraordinarily unfair, suspicious, hurtful, and just plain exhausting to deal with.

Later, the poster argues for engagement, since all people are worth listening to and dealing with. He positions himself, along with other Side B Christians, as complainers (i.e. trolls) and advocates for patient engagement of the type he sees advocated by his faith:

> However, I sometimes get the impression that a lot of us take this as a reason to write them off, and encourage others to do the same.... Where would be if God had decided to treat us as the trolls we so often are to him? Patience, love, and forgiveness are not optional extras for the Christian: they are the stuff on which we live.

Indeed, criticism of conservative Christian writers may be off-putting for new members of the group from conservative evangelical backgrounds, who may respond by leaving the group or staying silent.

In summary then, these minority-within-minority groups can build in-group solidarity, and even derive a certain sense of encouragement, from taking negative stances toward other larger-scaled and potentially threatening identity groups. Yet, for these Christians, taking negative stances toward other Christians, even if those Christians are insensitive to their particular minority group, is a problematic act. Such complaints about individual Christians, their local physical parishes or the Christian church as a whole are either rarely made, or contested, hedged and regretted when they are made.

Discussion

Virtual parishes built around minority identities exist precisely because of what physical parishes cannot provide. While detached from physical geography, virtual parishes thus owe their existence to users' physical parishes, and encourage and equip users to stay engaged with their faith in the context of those physical parishes. Taking into account the totality of how the Koryo-saram community, the Faith in the Academy group and the Side B group behave in their 'virtual parishes', two significant ramifications emerge for the study of minority groups' digital lives and of digital expression of Christianity.

Scale

Several of the dominant social actions taken through language noted in the 'Findings' section – offering specific encouragement, ventriloquating authoritative voices which lend an intersectional identity credence and constructing solidarity vis-à-vis threatening others – are relevant precisely because these are unavailable to smaller-scaled identities within physical parishes. As noted in the introduction, scale – consisting of both distribution and density – is an important consideration for analyzing language choices (e.g. Mortimer & Wortham, 2015; Wang et al., 2015). Due to sparse concentration and wide distribution, the chances are slim in the physical parish of encountering someone who can offer fitting encouragement, knows an appropriate authoritative figure or who fears the same other (those 'others' may indeed predominate within the physical parish). It makes sense then that the more small scaled an intersectional identity is, the more reliant users would be on the virtual parish to create a feeling of larger scale, with a higher number of instances of the needed action, occurring over a shorter period of reading and chatting, with a greater number of individuals. Indeed, the virtual parishes can be ranked in terms of how organized and active each of them were: Koryo-saram the least active, Faith in the Academy more and the Side B community the most active. Activity here is seen as synthesizing several factors – the degree to which all nominal participants were active contributors, and the relative frequency with which the participants tended to contribute.

It seems that the deeper a minority group is embedded within other nested layers of dominating identities, and the more vulnerable that group is to threats from those larger groups, the more significant a virtual parish can be for the sustenance and expression of faith. The desire to seek out other academics, or other Side B Christians, even if that can only be done virtually, is what initiated and sustains their better-organized virtual parishes. That the Koryo-saram network is less active and less organized may be because the Koryo-saram identity is less vulnerable, or because the physical parishes among which they are distributed tend to contain other Koryo-saram. Their unique multilingual repertoire also makes their intersectional identity an asset to Korea and to their employers, which may buffer some of that potential threat. Physical parishes may occasionally be culturally insensitive, but will not question their biographical identity claims on theological grounds. Christian academics are in some way vulnerable as a minority among academics for being Christian, and as a minority among Christians for being academics. Nevertheless, these two superset identities each bring a significant amount of social capital, especially in the American context. Yet, to the extent that belonging to the secular academy is seen as betraying Christian principles, or that having faith is inconsistent with modern education, the potential for threats remains. The Side B community – a minority of all LGBT, a minority of all Christians and even a minority of all LGBT Christians – has a very wide and very sparse distribution. Most Side B'ers do not have, or perhaps due to taboos not know of, other Side B Christians in their physical parishes. Whereas Side B'ers claim that the LGBT identity is biographical, some churches position that identity as subjectivity based, an intentional choice on their part. In the larger cultural debates about homosexuality, Side B Christians are sensitive to rhetorical threats, ranging from flaming to terminological insensitivity, from all of those superset groups. For a group with such a low density of distribution and many-layered nesting within potentially hostile superset groups, the virtual world becomes an encouraging safe space to negotiate the implications of their faith for their particular identity.

Charity

Another implication for the study of Christian online practice is the degree to which the contributions in these forums were governed by a sense of charity – forgiving intentional or unintentional insults and assuming good motives – especially toward physical parishes. Although larger-scaled groups of Christians such as Korean Protestants, evolution and climate change skeptics, or conservative evangelicals are potential sources of threat, such groups are not complained about or attacked as much as might be supposed. Also, while these virtual parishes only exist because of some unmet need, physical parishes were rarely blamed for not meeting those needs. Even among the Side B Christians, where frustrations with Christian teaching and practice were most frequently and clearly voiced, the local physical parish was largely spared from criticism. The frustration of blog writers or Facebook posters was nearly always aimed at the 'the (American) church' or meaning the aggregate disposition of all (Western) Christians, or else specific individuals whose influential voices were positioned as inaccurate and unrepresentative of Christian teaching. The physical parish, especially in the Side B blog, was upheld as the central organizing unit for Christian practice, with sanctification occurring via being challenged and changed by others of diverse backgrounds within local parishes. Side B Facebook or blog posts often seem to troubleshoot individuals' challenges within their own parishes, equipping them with strategies to transform their problematic experiences there. For example, one participant initially expressed sufficient frustration with his local parish to desire to leave:

UPDATE.

I am leaving my church.

I make this decision after much thought and prayer. It is no longer a safe space for me – and if I can't feel safe at church, where can I?

The response of the group to this post was sympathy, yet encouragement to work through the issues. Indeed, this poster decided not to leave the church in the end, and had transformative discussions with parish leadership. Indeed, restoration of strained relationships with friends and leaders in the local parishes was a common rhetorical theme. In the Side B blog, the authoritative posters there frequently celebrate the physical parish, to the point of encouraging Side B Christians to find in the parish the object of their desires for intimacy and love.

Although many types of diversity exist in physical parishes, parish members by definition identify with a single denomination or confession. Charity in the virtual parishes, however, also extended toward other Christian traditions and experiences, as a result of interactions held online. In the Side B Facebook group, posters expressed appreciation for encountering other Christian traditions deeply, as illustrated by the following post:

I know that many of us for various reasons have found (and needed!) cross-denominational support, so you may find [an article on Christian unity] encouraging. What gifts for the Christian life have you been pleased to find in a tradition other than your own?

Faith in the Academy members also appreciated learning about different parishes in their shared city through the contacts formed in the group. Interestingly, the fellowship engendered through sharing a different minority identity may lead Christians to their first intimate encounters with very different traditions, and charity toward a greater range of Christian biographies and subjectivities than those found within their own parishes.

Conclusion

In the 21st century, mobility has increased both the value placed on diversity and the self-selected homogenization of neighborhoods and community organizations. While parishes are more able than ever to draw people from great distances that meet a particular niche profile, some identity combinations remain so specific – such as the Koryo-saram, academic, and Side B communities – that physical parishes based solely around these identities are unlikely to appear. Indeed, while members of these minority groups clearly benefit from interacting with others 'just like me', these Christians also value fellowshipping with diverse believers in their physical parishes. That these physical parishes are not able to fully understand or meet the needs of each combination of minority identities, does not lead to indictment of these parishes. Virtual parishes take up a supplemental, not oppositional, role by offering solidarity, facilitating parachurch meetings, resourcing participants and providing tailored encouragement not found in the physical parishes. While members of minority groups do turn to digital communities to meet unmet needs, it seems that virtual parishes do not displace members' allegiance toward their physical parishes, but rather encourage charity toward, and the transformation of, participants lives within those physical and local communities of faith.

References

Bakhtin, M.N. (1981) *The Dialogic Imagination: Four Essays* (M. Holquist, ed.; C. Emerson and M. Holquist, trans). Austin, TX: University of Texas Press.
Bean, L. (2014) *The Politics of Evangelical Identity: Local Churches and Partisan Divides in the United States and Canada.* Princeton, NJ: Princeton University Press.
Berlinerblau, J. (2005) *The Secular Bible: Why Nonbelievers must Take Religion Seriously.* Cambridge: Cambridge University Press.
Blommaert, J., Westinen, E. and Leppänen, S. (2015) Further notes on sociolinguistic scales. *Intercultural Pragmatics,* 12 (1), 119–127.
Bozard, R.L. and Sanders, C.J. (2011) Helping Christian lesbian, gay, and bisexual clients recover religion as a source of strength: Developing a model for assessment and integration of religious identity in counseling. *Journal of LGBT Issues in Counseling* 5 (1), 47–74.
Brunn, S., Jeton, W. and Palmquist, B. (2015) Marketing religion and church shopping: Does one size fit all? In S.D. Brunn (ed.) *The Changing World Religion Map* (pp. 2285–2307). Dordrecht: Springer Science and Media.
Bucholtz, M. and Hall, K. (2004) Language and identity. In A. Duranti (ed.) *A Companion to Linguistic Anthropology* (pp. 369–394). Oxford: Basil Blackwell.

Campbell, H.A. (2005) Spiritualising the internet. uncovering discourses and narratives of religious internet usage. *Heidelberg Journal of Religions on the Internet* 1 (1). *http:// archiv.ub.uni-heidelberg.de/volltextserver/5824/1/Campbell4a.pdf*

Campbell, H.A. (2012) Understanding the relationship between religion online and offline in a networked society. *Journal of the American Academy of Religion* 80 (1), 64–93.

Canagarajah, S. (2002) Multilingual writers and the academic community: Towards a critical relationship. *Journal of English for Academic Purposes* 1 (1), 29–44.

Cimino, R. (2011) Neighborhoods, niches, and networks: The religious ecology of gentrification. *City & Community* 10 (2), 157–181.

Du Bois, J.W. (2007) The stance triangle. In R. Englebretson (ed.) *Stancetaking in Discourse: Subjectivity, Evaluation, Interaction* (pp. 139–182). Amsterdam: John Benjamins Publishing.

Edwards, K.L., Christerson, B. and Emerson, M.O. (2013) Race, religious organizations, and integration. *Annual Review of Sociology* 39, 211–228.

Ghaziani, A. (2015) 'Gay Enclaves Face Prospect of being Passé': How assimilation affects the spatial expressions of sexuality in the United States. *International Journal of Urban and Regional Research*. doi: 10.1111/1468-2427.12209

Gross, M. and Yip, A.K. (2010) Living spirituality and sexuality: A comparison of lesbian, gay and bisexual Christians in France and Britain. *Social Compass* 57 (1), 40–59.

Gross, M. (2008) To be Christian and homosexual: From shame to identity-based claims. *Nova Religio* 11 (4), 77–101.

Hall, D. (2015) Individual choices revisited: Non-heterosexual Christians in Poland. *Social Compass* 62 (2), 212–224.

Hammett, J.S. (2000) How church and parachurch should relate: Arguments for a servant-partnership model. *Missiology: An International Review* 28 (2), 199–207.

Kim, G. (2003) Koryo Saram, or Koreans of the former Soviet Union: In the past and present. *Amerasia Journal* 29 (3), 23–29.

Kinnaman, D. (2011) *You Lost Me: Why Young Christians are Leaving Church... and Rethinking Faith*. Grand Rapids, MI: Baker Books.

Kubicek, K., McDavitt, B., Carpineto, J., Weiss, G., Iverson, E.F. and Kipke, M.D. (2009) "God made me gay for a reason": Young men who have sex with men's resiliency in resolving internalized homophobia from religious sources. *Journal of Adolescent Research* 24 (5), 601–633.

Lewis, C.S. (2001) *The Screwtape Letters*. Grand Rapids, MI: Zondervan (original work published 1942).

Marti, G. (2009) *A Mosaic of Believers: Diversity and Innovation in a Multiethnic Church*. Bloomington: Indiana University Press.

Meyerhoff, M. and Niedzielski, N. (1994) Resistance to creolization: An interpersonal and intergroup account. *Language & Communication* 14 (4), 313–330.

Mortimer, K.S. and Wortham, S. (2015) Analyzing language policy and social identification across heterogeneous scales. *Annual Review of Applied Linguistics* 35, 160–172.

Suler, J. (2004) The online disinhibition effect. *CyberPsychology & Behavior* 7 (3), 321–326.

Taylor, Y., Falconer, E. and Snowdon, R. (2014) Queer youth, Facebook and faith: Facebook methodologies and online identities. *New Media & Society* 16 (7), 1138–1153.

Vervoort, M., Dagevos, J. and Flap, H. (2012) Ethnic concentration in the neighbourhood and majority and minority language: A study of first and second-generation immigrants. *Social Science Research* 41 (3), 555–569.

Wang, X., Spotti, M., Juffermans, K., *et al.* (2014) Globalization in the margins: Toward a re-evalution of language and mobility. *Applied Linguistics Review* 5 (1), 23–44.

Whitehead, D. (2015) The evidence of things unseen: Authenticity and fraud in the Christian mommy blogosphere. *Journal of the American Academy of Religion* 83 (1), 120–150.

Zhao, S., Grasmuck, S. and Martin, J. (2008) Identity construction on Facebook: Digital empowerment in anchored relationships. *Computers in Human Behavior* 24 (5), 1816–1836.

3 Facebook: A Medium for the Language Planning of Migrant Churches

Ana Souza

Ethnic churches have increasingly adopted the use of technology as part of their services and as a way of maintaining transnational links (Oosterbaan, 2011). Consequently, the virtual world has impacted on the religious practices of migrants (Vásquéz, 2010) and on their use of languages. Technology may contribute to the transmission of religious beliefs. However, technology can threaten the transmission of minority languages (Annamalai, 2006). This chapter adopts a micro-perspective to the analysis of language planning (Souza *et al.*, 2012) in faith settings in London. The interface of language, religion and technology is explored through the examination of computer-mediated communication. The discussions are illustrated with examples from Pentecostalism, one of the most common religions among Brazilian migrants. It is argued that Pentecostal groups use technology to develop virtual transnational networks that will support their language ideologies in relation to the transmission of religious beliefs.

Introduction

The growing number of minority groups in a variety of countries has led to the presence of diverse minority media, which include radio, newspapers, magazines, TV, cinema and the internet. This phenomenon has been the focus of attention of minority language media (MLM) researchers since the 1980s (Browne & Uribe-Jongbloed, 2013), whose concerns are with how media can be used to support the use of minority languages (Cormack, 2013). Nevertheless, the internet and social networking sites (SNSs) have been seen as much an opportunity as a threat to language maintenance (Cunliffe *et al.*, 2013). Although the social and participatory-democratic potentialities of these sites have been seen as positive to the communication of minority groups, the basic concepts of the mass communication paradigm are still very much alive (Carpentier, 2009 in Gruffydd Jones, 2013: 70). In other words, it is still possible to experience exposure of controlled content to a large audience in a unidirectional and unmediated way with the aim of creating uniform messages.

Despite this mass communication paradigm, media have played a limited role in 'top-down' language plans (i.e. plans made by governments that are imposed on a group), as they are usually beyond the influence of language planners (Cormack, 2013). Instead, new media – the internet and SNSs, in particular – allow their users to develop their own activities, and thus be involved in 'bottom-up' initiatives (Cormack, 2013). In fact, new media have been reported to generate new forms of social gatherings, including religious ones (Sanchotene, 2011). It is now possible to 'attend' religious services online, take part in praying sessions via the internet, watch previously recorded religious programmes on websites and interact with fellow religious members via SNSs.

This situation has led some researchers (e.g. Sanchotene, 2011) to believe that the internet may lead to a new relationship with religion – one in which face-to-face contact is replaced by the computer screen. Other researchers (e.g. Lundby, 2011) see the internet as an online space that interacts with and influences offline spaces. An illustration of this is Souza's (2014) study of 'Kardecists' (Christians who also believe in reincarnation and spirit mediumship) in London. The use of the internet by the teachers who deliver the faith lessons to the Kardecist children allows for their connection with offline spaces as well as for their development of virtual transnational networks. More specifically, the Kardecist teachers in England decided to adopt English as the language of interaction in their lessons. With this purpose in mind, they developed online links with an organisation in Brazil to access their lesson plans in Portuguese. These plans are translated into English by the Brazilian teachers in the UK, who then post the plans online to be accessed by Kardecist Brazilian migrants in other countries. The teachers in the UK also meet face to face to discuss these plans and their lessons. In other words, the online networks support the language ideologies of the Brazilian leaders in relation to the transmission of their religious beliefs offline.

In this chapter, I focus on a group of Brazilian Pentecostal migrant churches and their use of Facebook as a medium for language planning, i.e. the deliberate choice of language to be established as the language a group of speakers should adopt. I draw on Blommaert's (2007) notion of sociolinguistic scales, which presents language planning and policy (LPP) as being part of a multilayered and dynamic process that varies according to time and space. More specifically, I explore the language choices made by a group of faith leaders in Portugal, Italy, the USA and the UK when posting on their Facebook pages in contrast to the choices made by their followers. I argue that the linguistic positions adopted by faith leaders on Facebook impact the level of online interaction of their migrant members. I conclude by pointing out that the most successful migrant churches, in engaging their followers online, are the ones that adopt a flexible approach to LPP.

Theoretical Framework

LPP was developed as a field of study after the Second World War when nations were being (re)built (Spolsky, 2012). Ricento (2000) discusses LPP as having three phases: structuralism in the 1960s, post-structuralism in the 1970s/1980s and critical sociolinguistics in the 1990s. In the structuralism period, there were two types of LPP: corpus (developing and manipulating language forms such as orthography and grammar) and status (allocating functions and uses for specific languages). Corpus and status types of LPP were still present in the post-structuralism period. These two types of LPP were based on the premise that linguistic diversity is a problem (Mühlhäusler, 1996 in Hornberger, 2002: 32). However, three important developments happened in this second period of LPP: its focus moved to contexts, to acquisition type of planning (language teaching to increase the number of speakers of a specific language) and to the sociopolitical and ideological nature of LPP (Ricento, 2000). These developments led to the critical sociolinguistics period, when the interests of dominant social groups were acknowledged and the social inequality these interests caused was at the centre of the studies. As summarised in Hornberger (2002: 35), the 'language planning field ... moved from a focus on problem-solving through a concern for access and into an emphasis on linguistic human rights'.

The layers of LPP

Johnson and Ricento (2013) extend the analysis of LPP to the 21st century and refer to it as the ethnography of language policy phase. The ethnography of language policy is a method for examining the agents, contexts and processes across multiple layers of LPP (Hornberger & Johnson, 2007). These multiple layers have been referred to as the 'onion layers' by Ricento and Hornberger (1996) and are the (1) macro, (2) meso and (3) micro layers. The macro layer refers to the political processes of a nation. The meso layer relates to the institutions present in a society. The micro layer concerns the language negotiations at the interpersonal level of a group of people. This metaphor of the 'onion layers' is of much importance in highlighting the new LPP perspective on communities, in other words, on bottom-up planning (Ferguson, 2010). The macro and micro perspectives are combined in the ethnography of language policy studies, which provide a balance between policy power and interpretative agency (Johnson & Ricento, 2013).

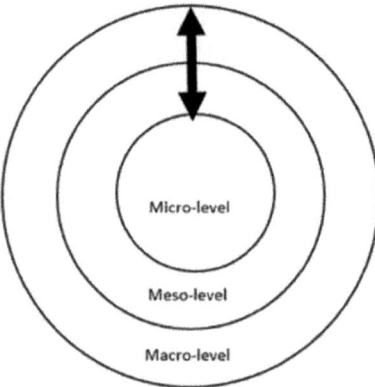

Figure 3.1 Visualising the macro, meso and micro levels of LPP

If we were to visualise the hierarchical relationship between the levels of LPP through the 'onion' metaphor, we would have an image such as Figure 3.1.

The meso layer of LPP

One example regarding the importance of the meso layer of LPP is Musk's (2010) study on bilingual education in Wales. This study shows how schools have played a key role in the revitalisation of Welsh and in the increasing number of Welsh–English bilingual individuals. Musk (2010) draws on Butler's (1990) notion of the performative to gender to claim that bilingualism is

> a social construction, that is, a dynamic, non-essentialist category that is produced by means of repeated discursive acts ... [i.e.] discourse relating to the phenomenon [of bilingualism], but also as the everyday situated practices (interactions) of bilinguals ... [and which have] a reflexive relationship ... [with each other]. (Musk, 2010: 42)

Musk (2010) examines three sets of data: a Welsh language policy document (macro level), a video recording of one of the schools' open evenings (meso level) and video recordings of Year 12 pupils' focus groups (micro level), of which only one is examined in his article. Musk focuses his discussions on the linguistic diversity discourse, which relates to the promotion of the maintenance of one's diverse linguistic heritage. The core of the linguistic diversity discourse in the language policy document analysed is a call for a commitment to Welsh and to bilingualism. This same discourse is observed in the school when the principal addresses prospective pupils and their parents in their open evening and makes reference to its commitment to the development of the Welsh language. Musk highlights, however, that there are some differences in the way that this commitment is framed at different levels. At the meso level, the promotion of linguistic diversity is recontextualised to the relevance of this diversity to examination results, and thus, its commodification. Recontextualisation is also reported by Musk in relation to the pupils' focus groups. One of the pupils, for example, appropriates the school's linguistic rights discourse in protection of Welsh to argue for her preference for speaking English. Nevertheless, speaking Welsh is recognised as one of the school's core values. Musk also points out that the majority of the pupils use both Welsh and English outside school, and thus, the school, i.e. the meso level, has had an impact on the language practices of Welsh society, i.e. on the micro level. In other words, the introduction of the meso level highlights the need to go beyond the static macro–micro dichotomy traditionally adopted in the studies of LPP.

The LPP layers in continua

Hornberger (2002) draws on the continua of biliteracy, a dynamic framework, in her study of educational language policy implementation in South Africa and in Bolivia in the 1990s. This framework refers to language planning, research and teaching biliteracy (i.e. communication in/about writing in two or more languages) in multilingual settings. Within this framework, the biliteracy development (reception–production, oral–written, first language–second language [L1–L2]) is described along intersecting continua in relation to contexts (micro–macro levels, oral–literate, bi/multilingual–monolingual), media (simultaneous–successive exposure, dissimilar–similar structure, divergent–convergent scripts) and content (minority–majority perspectives and experiences, vernacular–literary styles and genres, contextualised–decontextualised language text).

As explained by Hornberger (1989: 28), 'the implications of [this] model of biliteracy ... are that the more contexts of [the] learning allow [learners] to draw on all points of the continua, the greater are the chances for their full biliterate development'. Hornberger (2002: 40), however, warns us about the fact 'that societal power relationships tend to favor the macro, literate, and monolingual ends of the context continua'. Both institutions and individuals are affected by these relations of power. The Bolivian and the South African parents in Hornberger's (2002) study, for example, were clearly influenced by their experiences of linguistic imperialism with Spanish and English, respectively. As a consequence, they questioned the multilingual language policy of their countries. In other words, 'the zeal of educators and policy makers for teaching children literacy on the foundation of a language they already speak appears to be at odds with a popular demand for the language of power' (Hornberger, 2002: 38).

However, the challenge of negotiating across multiple languages is not restricted to the context of education. Therefore, I turn to another framework that could be applied to a variety of contexts, the sociolinguistic scales.

LPP and the notion of 'scales'

The macro–micro dichotomy has been more strongly criticised by Blommaert (2007), who argues that it should be replaced by the notion of 'scales'. According to him, this notion

> suggests that processes of distribution and flow [horizontal metaphor] are accompanied by processes of hierarchical ordering [vertical metaphor], in which different phenomena are not juxtaposed [placed or dealt with close together for contrasting effect], but layered and distinguished as to the scale on which they operate.... (Blommaert, 2007: 1)

Furthermore, Blommaert (2007) introduces the concept of TimeSpace to sociolinguistic scales, i.e. to the processes of distribution and flow. Time and space are here presented as a 'single dimension' (TimeSpace), as 'every social event develops simultaneously in space and in time, often in multiple imagined spaces and time frames' (Wallerstein, 1997 in Blommaert, 2007: 5). Blommaert *et al.* (2006) illustrate the effect of the TimeSpace scale in sociolinguistic processes with a specific focus on the assessment of one's linguistic competence. They show how the lack of recognition of migrant children's languages leads them to stop being seen as holding complex literacy skills but rather being seen as illiterate in Belgian immersion schools, where the only recognised language is Dutch.

The micro (local) and the macro (global) layers are presented by Blommaert (2007: 1) as scales in the extremes of a continuum on which 'social events and processes move and develop'. A very relevant point made by Blommaert (2007: 1–2) is that 'interactions between the different scales [are] core feature[s] of understanding ... events and processes'. In other words, his notion of scales tries to make explicit the links between the macro and micro levels of sociolinguistics. In Blommaert's (2007: 4) own words 'the introduction of "scale" does not reject horizontal images of space; it complements them with a vertical dimension of hierarchical ordering and power differentiation'. Based on this, I found it useful to picture these images in a diagram (see Figure 3.2).

Hult (2010) has considered the application of sociolinguistic scales to be relevant for the ecological understanding of the relationships between LPP and the language use of individuals. That is, the conceptual orientation of speakers' use of language is situated in social contexts and can be better understood by reference to time and space. In order to justify his position, Hult (2010) uses a microscope analogy:

> What are often theorized as 'layers' are essentially the result of an analytical lens ... One may choose a specific location in a linguistic ecosphere on which to train one's microscope. The 'level' we see, then, is a question of the power of magnification. Once we choose a power

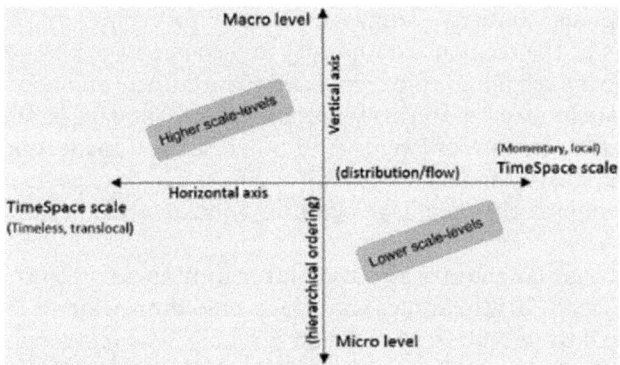

Figure 3.2 Picturing the sociolinguistic scales continuum (based on Blommaert, 2007)

of magnification, we may then focus our view of an object of study to see different features of it, and more or less of the surrounding context, depending on the focus [Garner 2004: 202]. Later, we may also change the power of magnification to visualize a different perspective. What is being examined, then, are not distinct layers but what Blommaert [2007] refers to as 'scale'. (Hult, 2010: 14)

As previously described, scale is a sociolinguistic construct that refers to the fluid and dynamic nature of relationships among discourse processes across dimensions of social organisation as situated in time and space. Consequently, scales are interdependent and connected to each other by the people and the discourses that move between them (Hult, 2010). It means that although language policies are situated in a particular TimeSpace scale, they influence and are influenced by other TimeSpace scales. Hult (2010) illustrates this by referring to his 2007 study of Swedish educational language policy. He shares an extract of the introduction to a preservice English as a foreign language course being made by its instructor. The data show that the instructor moves back and forth between focusing on the trainees' future internship and referring to the national Swedish curriculum. This interaction indicates that discourses from the macro scale (the policy document) need to be made relevant at the micro scale (the internship) (Hult, 2010). In spite of my use of the words 'macro' and 'micro' to refer to Hult's work, it should be noted that – as scales – they are not perceived as discrete layers. On the contrary, the interconnectedness of the macro and the micro dimensions of social contexts is highlighted by the notion of scales.

LPP, religion and scales

Although the four examples of LPP negotiations in the previous section refer to educational contexts, the concepts discussed are also relevant to religious ones. As discussed by Liddicoat (2012b),

[i]n religious contexts, language is used for communication among members of the religious community and so language plays a significant role in how a religion is communicated to its faithful and how the faithful participate in religion. In this sense, the use of language in religious life is analogous to the use of language in other, secular, institutions. For such uses of language, language planning activities can be expected to resemble those found in other language planning contexts. (Liddicoat, 2012b: 121)

Considering that language is also used for communication to and about the divine, Liddicoat (2012b) argues that the sacred dimension of language use may also need to be addressed in studies of LPP. This relationship between language and religion has been acknowledged by the sociology of language and religion (SLR), a subfield of sociolinguistics which developed in the 21st century (Darquennes & Vandenbussche, 2011) and which was embraced by studies such as those presented in this chapter.

Woods (2006) explored LPP at both the meso and the micro layers in her study of Christian denominations of varied linguistic and ethnic backgrounds in Melbourne, Australia. She observed that the role of language was valued differently by each of these denominations, and so, she developed a framework to explore the links between their language ideologies (i.e. beliefs and values in relation to language) and language practices (i.e. patterns of language use), the 'language–religion ideology' (LRI) continuum. As she explains, 'the formation of the language ideology of a denomination is largely a product of its theological orientation' (Woods, 2006: 201). It means that some denominations may allocate a special language to communicate with/about God, for viewing this relationship as a very special one, whereas other denominations may emphasise a personal relationship with God and thus create space for the use of vernacular languages in their worship. These perspectives of language have been named as 'a sacral view' (i.e. the use of a language is a sacred act) and as 'a comprehensible view' (i.e. the use of a language is a communication act) by Liddicoat (2012b). Migrant churches are also affected by the cultural values that their communities attach to their heritage languages. The theological orientations and cultural values of migrant churches may lead to some internal conflicts in relation to LPP. This is illustrated by Woods (2006) in her example of a Latvian Lutheran congregation. The minister considered English appropriate to be used with the youth as part of the services, which reflects the Lutheran position of individual access to their Scriptures. However, the value posed by the older members of the congregation on the Latvian language – due to their objective to preserve their heritage – is in conflict with the LPP adopted by their minister.

It is possible, however, to find examples of churches where there is a match between the position of a congregation and that of their leaders. An illustration of this is the case of the Brazilian Catholic setting in London, UK, reported in Souza et al.'s (2012) study. The theological orientation of this faith setting was to support migrants abroad and to offer them support in Portuguese, their mother tongue. This concern affected the religious services offered to the adult congregation, such as Mass (i.e. the Catholic communal worship) and those offered to their children, such as the catechism (i.e. faith lessons that introduce children to the Catholic sacraments). According to the priest, the decision to deliver faith lessons in Portuguese was a consequence of the importance that the Brazilian Catholic chaplaincy in London gave to language maintenance and to the children's emotional and cultural links to Brazil. In other words, the study found that the religious, ethnic and linguistic dimensions of identity were reinforced in the Brazilian Catholic lessons. Consequently, a new framework was suggested for the studies of LPP, which could be applied to both educational and religious contexts, the religion–ethnicity–language (REL) triangle (Souza, 2015). In this framework, each of the three aspects of identity (i.e. religion, ethnicity and language) is placed at one of the angles of the triangle with a continuum moving inwards. A move towards the inner extremes of each of the continua represents weaker identity links with that aspect of their identity, whereas a move outwards means stronger links.

Another framework that can help the understanding of the matches and mismatches between theological and cultural orientations of LPP in religious contexts is Blommaert's (2007) notion of scales, which was discussed in the previous section. This notion developed from his work with colleagues on criticisms of the accounts of linguistic and communicative competence in relation to multilingualism (Blommaert et al., 2005). As Blommaert et al. (2005: 200) put it, '… space should be seen in connection to scaling processes; movements across space involve movements across scales of social structure having indexical value and thus providing meaning to individual, situated acts'. Blommaert et al. (2005) also remind us that movements across spaces take place with material and symbolic attributes and features – spaces are not equal, they stand with each other in relationships of power and value.

Having introduced the theoretical background to the discussions in this chapter, I present the study on which they are based.

The Study: Participants, Methodology and Data Analysis

The study reported in this chapter examines the attempts of four Brazilian Pentecostal migrant churches to use Facebook as a way to introduce the use of specific languages to their followers.

Participants

The Brazilian Pentecostal migrant church reported in this chapter has five international branches and seven sub-branches in European countries and the USA. The international branches and sub-branches are linked to the main church via bidirectional visits between leaders in Brazil and leaders abroad, the availability of a website which brings together all the national and international (sub-)branches of the church, a weekly television programme broadcast on a Brazilian satellite channel and a digital radio station that plays Brazilian Gospel songs and can be accessed online.

Although the members of the international (sub-)branches are mostly Brazilians, people of a variety of cultural and linguistic backgrounds have joined them throughout the years. Therefore, besides the weekly face-to-face services offered in Portuguese, special services are also offered in the local languages. In addition, four of the five international branches have created a Facebook page and thus are the focus of the study reported in this chapter.

Methodology

The data for the discussions in this chapter were extracted directly from the churches' Facebook pages online. As this medium is in the public domain, there was no direct interaction with the participants or reliance on self-reporting instruments. Zimmer (2010) seems to question the ethics of such a procedure. However, his reflections are on closed Facebook pages, i.e. groups to which participants have to be invited or have to request authorisation to join the page. The information shared in closed groups is not available to the public, which is not the case for the pages analysed in this chapter.

Nevertheless, care has been taken to avoid presenting details that could lead to the identification of the churches or of anyone who contributed to the Facebook pages referenced in this study. As a consequence, the name of churches, their Facebook page address, images of posts considered for analysis and the names of church followers are not displayed. This procedure ensured that personal information was not accessed improperly, no unauthorised use was made of personal information, no unauthorised secondary use of information was made and that there were no errors in relation to personal information shared about the participants (in this case the Facebook users). In other words, none of the four error violations that can occur when personal information is accessed through conducting online research, as raised by Smith *et al.* (1996 in Zimmer, 2010), took place in this study of Brazilian Pentecostal migrant churches.

Data analysis

The Facebook page of each of the four Brazilian Pentecostal migrant churches was visited in April 2015 with the purpose of collecting messages posted by both faith leaders and their followers. All the posts published in the year of data collection, i.e. between January and April 2015, and the year each Facebook page was created were copied and pasted using a computer print-screen key. A word file was created for each of the churches with these print-screened images of the posts. Due to the above-mentioned ethical considerations, the print-screened posts are not shared in this chapter. Nevertheless, the posts were used for the analysis stage and grouped according to their content, i.e. promotion of local events, promotion of events in Brazil, witnessing (sharing one's faith in Christ), evangelisation (the proclamation of the good news of Jesus Christ) and praising (giving thanks to God). Notes were then taken on the languages used in each of the posts (i.e. Portuguese, English and/or Italian) and by whom (i.e. leaders or followers). In this way, an understanding of which languages and of how they are used online could be drawn. This analysis of the posts was the basis for writing a description of the language use for each of the four Facebook pages. The notion of sociolinguistic scales discussed in the section 'Theoretical Framework' was then used to reflect on the impact that time and space had on the use of language in the different online spaces.

Facebook and Brazilian Pentecostal Migrant Churches

Facebook is a type of SNS which allows the creation of virtual communities. A Facebook page can be created online and for free by anyone who wants to promote anything that he or she may find interesting, including religion. Facebook is the largest social network in the world with 1425 billion users as of March 2015.[1] Its growth around the world has made language an important factor for reaching local and global audiences.[2] Portuguese is its third most used language, and Brazil,[3] which is expected to reach 90.61 million users in 2018,[4] is the second country with most users.[5] The three largest Facebook pages in Brazil are linked to Catholicism.[6] However, these statistics change if the focus is on Brazilian churches that offer online services to followers who have migrated to other countries. In this case, it is the Pentecostal migrant churches that have a larger number of Facebook pages, and thus, the largest number of users.

The Brazilian Pentecostal church in question has its main branch in Brazil and further branches in five different countries. Three of these branches are in three Swiss cities, one branch in Portugal, four branches in four cities in Italy, one in the USA and one in the UK. The latter four countries, i.e. Portugal, Italy, the USA and the UK, have a Facebook page. The deliberate choice of language made by the Brazilian Pentecostal leaders of each of these four Facebook pages is analysed in the next sections. Although the data were in the public domain, care was taken to protect the identity of the churches, their leaders and followers, as detailed in the 'Methodology' section. As a consequence, the analysis that follows can only present a general picture of the language choices being made. Nevertheless, the findings point to interesting patterns which could benefit from further and more detailed analysis, if authorisation to do so is provided by the participant Facebook users.

The Brazilian Pentecostal church in Portugal

Portugal is the European country with the largest number of Brazilian migrants (Ministério das Relações Exteriores [MRE], 2012). The high presence of Brazilians in Portugal could be related to the colonial links between the two countries, which has led to blood relations and sharing the same language, albeit with different regional and national varieties.

However, as mentioned above, the Brazilian Pentecostal church has just one branch in Portugal. Their geographical concentration in one city might explain the late creation of this page, March 2015, in relation to the other three international branches. The fact that all their followers live in the same city where the branch holds its face-to-face activities may have delayed a need to communicate via Facebook. Nevertheless, within a month of being created, this Facebook page had 147 likes.

Because Portuguese is the national language of both Brazil and Portugal, language planning does not seem to be an issue for the international branch in the latter lusophone country. All the posts in this Facebook page are in Portuguese, as expected. However, it is clear that the leaders and their followers use the Brazilian variety of the Portuguese language in all their posts. This language use may reflect the profile of the 'church goers' at the time the data were collected. It will be interesting to observe how this profile changes as the church grows and how this growth impacts on the use of other varieties of Portuguese online, especially the European one, i.e. the national variety in Portugal.

The Brazilian Pentecostal church in Italy

Italy is the fifth-ranked European country in number of Brazilian migrants (MRE, 2012). Considering that Brazil has a large number of Italian descendants, which has led to a number of Brazilian nationals holding Italian passports, it is initially surprising that there are not more Brazilian migrants in this country. One possible explanation for this is that Italy is used by many Brazilian nationals as an entry door to Europe. This has been noticed in relation to the UK, where a significant number of Brazilian migrants are holders of Italian passports (Souza & Evans, 2015).

Nevertheless, the Facebook page of the Italian branch was the first one to be created in 2008, and had 4200 likes by the end of April 2015. The high number of followers in Italy may be related to the fact that there are sub-branches of the church in four different Italian cities. In other words, the geographical distance between the four branches might encourage their followers to communicate via Facebook.

The faith leaders[7] in the Italian Facebook page are seen using mostly Italian in their posts. Original messages in Portuguese (henceforth, used to refer to Brazilian Portuguese) are shared but with translations into Italian. However, there are no translations from Italian into Portuguese. One could say that the similarities between Portuguese and Italian, especially in writing, allow Brazilians to understand the messages without the need for translation, independently of their competence in Italian. Nevertheless, this fact appears to indicate that the language planning choices are favouring the non-Brazilian followers as a consequence of the church's possible intention of recruiting more members in their host country. This is a theological and a language ideology that has been demonstrated by pastors of the branch in England during semistructured interviews in a previous study (see Souza *et al.*, 2012).

The posts by the leaders refer to marketing events in Brazil and in Italy, the act of witnessing and religious messages (both evangelising and praising). There are also videos with subtitles in Italian and Twitter messages from the pastor in Brazil who writes in Portuguese and has his messages translated into Italian by the leaders in Italy. What happens, however, when the language of the host country differs more widely from Portuguese? In order to answer this question, I move on to the Facebook page of the branch in the USA.

The Brazilian Pentecostal church in the USA

The USA has the largest number of Brazilians outside of Brazil (MRE, 2012). Not surprisingly, the Brazilian Pentecostal migrant church has a branch in the USA as well as a Facebook page – created only a year before the one in Portugal, in 2014. The faith leaders in this branch mainly display messages in English about local events in the USA with no translation into Portuguese. However, all the messages about events in Brazil and praising are presented in Portuguese with translations into English. These are usually posts that contain pictures and cannot be changed. Therefore, translation is added to the post. In contrast, most of the comments by the followers are in Portuguese and the few comments in English tend to be set phrases such as the wish of a Happy New Year. These are phrases that may be used in Brazil too due to the influence of American and British movies and music on Brazilian society. In turn, the evangelisation messages are shared in two separate posts, one in English and one in Portuguese.

As indicated above, English is being used as a lingua franca, i.e. it is being used for communication between speakers who do not have English as their first language (Seidlhofer, 2005). This use of English is adopted by the leaders of the Pentecostal group in the USA as a strategy to reach out for local followers, be they American or migrants of other cultural and linguistic backgrounds. A similar strategy is adopted by the leaders in the UK, who also adopt the use of English as a global language, i.e. a common language that enables people from diverse linguistic backgrounds to communicate. The same can be said about the leaders in Italy, who adopt a standard variety of Italian over the various dialects used in that country to communicate with Italian and non-Italian followers.

As in the Portuguese branch of the Brazilian Pentecostal migrant church, videos are also shared via Facebook. The sharing of videos is a common practice of this medium. In the context of religion, their use may have a special meaning: videos may be part of what has been labelled as the 'faith of performance' (*a fé do espetáculo*) (see Sanchotene, 2011). As such, the experience is more relevant than the words, thus the videos of praising activities are originally in Portuguese and are not translated into English. It seems important to point out that these videos are uploaded by the church followers, not the leaders. The videos posted by the faith leaders relate to their work of evangelisation on the streets of the USA. As the aim of these activities is to convert the locals, English is the language used on the banners and in the communication between church followers and the people they address in the videos.

The Brazilian Pentecostal church in the UK

Despite being located in only one city, the branch in the UK seems to be aware of the fact that this country is third in its number of Brazilian migrants in Europe (MRE, 2012). It is also known that this group of migrants is spread all over the capital city and other cities in the four corners of the country (Souza & Evans, 2015). This context justifies the use of social media to reach a large number of (possible) followers. As a consequence, this branch was the second one to create a Facebook page in 2012 and had 9129 likes at the time the data were collected for this study.

The leaders of this branch tend to post information to promote local events in English. English is certainly used to promote events targeted at locals and youth. This choice of language seems to relate to possible changes in the linguistic characteristics of their followers. Intergenerational shift to the language of the host community is common as a consequence of migration and may affect the language choices of religious groups (Liddicoat, 2012b). This language choice may also be related to the church's intention of attracting non-Portuguese-speaking youth to their activities. Indeed, both groups seem to be the focus of these posts.

The posting of events in English, the local language, by the leaders is commonly followed by comments in Portuguese by the 'church goers'. The same phenomenon takes place in the branches in the USA. Videos about events which took place in Brazil are also shared by the followers with no translation to English, as is done by the branch in the USA. However, translations from Portuguese into English are commonly used by the leaders for posts on evangelisation and most recently for messages originally tweeted in Portuguese by the pastor in Brazil.

There are three occasions when the leaders post in Portuguese only. The first is when the messages in Portuguese can be understood by the non-Portuguese speakers. For example, when mentioning the name of a famous Brazilian gospel singer with the word *Grammy* in block letters to share the information that they had won this prize. The second occasion is to post short messages to praise the Lord. The choice of languages on this occasion appears to relate to the audience. This relationship between the language in the post and the targeted audience is confirmed by posts that are written in English inviting their followers to attend services in English and in Portuguese to announce the services in Portuguese. The third occasion in which the faith leaders post in Portuguese is when witnessing.

Discussion and Conclusion

Bell's (1984: 159) Audience Design Framework for explaining style variation 'assumes that ... speakers take most account of hearers in designing their talk ... [and thus] design their style for their audience'. In fact, Tagg and Seargeant (2013) have considered the application of Bell's framework, which was designed for oral public broadcast, to online contexts such as Facebook, where users interact via a written medium. Tagg and Seargeant (2013: 161) 'argue that audience design strategies are crucial for SNS users as they seek to target individuals and communities from within the wider audience'.

Tagg and Seargeant's (2013) modified version of the Audience Design Framework could be used for the analysis of the data presented in this chapter. This type of analysis would highlight the intentions of the posters of the messages in relation to what they knew and/or imagined about their audience (i.e. addressee, active friends, wider friends and the internet as a whole). Being aware of these intentions and knowledge is indeed relevant in understanding how multilingual users shape their online choice of languages in relation to how they perceive their Facebook audience, as advocated by Tagg and Seargeant (2013). However, the Brazilian Pentecostal leaders reported in this chapter are not only making choices about their own language use. These leaders also appear to be making choices about the languages they would like to see being adopted online and offline by their followers too. In other words, they are making LPP choices. As such, I argue for the application of Blommaert's (2007) notion of sociolinguistic scales to the use that these leaders of migrant churches are making of Facebook. As mentioned above, sociolinguistic scales allow for the understanding of the relationships between LPP and the language use of individuals (Hult, 2010).

The data collected on the Facebook pages of the four Brazilian Pentecostal migrant churches show that this medium is used for five purposes: to promote the local events organised by their international branches; to promote the events hosted by the main branch in Brazil; to share their faith in Christ (witnessing); to proclaim the good news of Jesus Christ (evangelisation); and to give thanks to God (praising). Despite the links that the migrant churches have with Brazil, the languages they use to perform these different communicative events on Facebook are not the same (see Table 3.1). Portugal, sharing its official language with Brazil, performs all five types of communication in Portuguese. In other words, the national language policy of the two countries is transferred to the online interactions of the church leaders and their followers in spite of their different language varieties. The same does not apply to the other three churches.

In examining the choice of languages made by the branches in Italy, the USA and the UK in the promotion of the events they organise locally, three different patterns of language use is noticed. The branch in Italy uses Italian in all of its posts and both languages in some of them. The branch in the USA previously posted information about their local events in Portuguese only. In January 2015, however, they started to publish the same events with two versions – one post in Portuguese and another in English. The branch in the UK uses language in a more complex way. The posts that target Portuguese speakers are in Portuguese and the posts that target non-Portuguese speakers are in English. Some of the posts even make use of translanguaging (i.e. drawing on more than one language to transmit a message [Blackledge & Creese, 2010]) in this case, Portuguese and English. More specifically, it is their common practice for the posts to present the name of an event in English and the event details in Portuguese. The use of numbers for dates and times supports the understanding of the message by the non-Portuguese speakers. In addition, images are also used to reinforce the ideas that are being conveyed. This practice calls attention to the multimodality involved in the interactions taking place in SNS. A post for the promotion of a concert, for example, showed the image of a singer holding his guitar, standing in front of a microphone.

Events hosted by the main branch in Brazil are promoted on the Facebook pages of its international branches before and after they take place. The pre-event posts are disseminated with images of the flyers with details about the date, time and venue for the event on all four Facebook pages. The original messages are presented in Portuguese with translations into Italian on the Facebook page of the branch in Italy and into English for publication on the Facebook pages in the USA and the UK. The post-event information is shared via the use of videos that are uploaded onto the branches' Facebook pages. All the videos are originally recorded in

Table 3.1 Facebook language use of four Brazilian Pentecostal migrant churches

		Local events	Events in Brazil	Witnessing	Evangelisation	Praising
Portugal	Leaders	Posts/Videos in Brazilian Portuguese	Posts in Brazilian Portuguese	–	Posts in Brazilian Portuguese Twitter: Brazilian Portuguese	Posts in Brazilian Portuguese Videos in Brazilian Portuguese
	Followers	Likes and shares Comments in Brazilian Portuguese	Likes and shares	Posts in Brazilian Portuguese Likes Comments in Brazilian Portuguese	Likes and shares Comments in Brazilian Portuguese	Likes and shares Comments in Brazilian Portuguese
Italy	Leaders	Posts in Italian Some posts in Italian and Portuguese	Posts in Brazilian Portuguese with translations into Italian Videos with subtitles in Italian	Posts in Italian Videos with subtitles in Italian	Videos with subtitles in Italian Posts in Italian Twitter: Brazilian Portuguese with translations into Italian	Videos with subtitles in Italian Twitter/Posts: Brazilian Portuguese with translations into Italian
	Followers	Likes and shares	Posts in Brazilian Portuguese with translations into Italian Likes and shares	Few posts in Italian Videos with subtitles in Italian Likes	Likes and shares Rare comments, mostly in Italian	Likes and shares Rare comments, mostly in Italian

(Continued)

Table 3.1 (Continued)

		Local events	Events in Brazil	Witnessing	Evangelisation	Praising
USA	Leaders	Posts in English, mainly. Some posts in Brazilian Portuguese with translation in English in a separate post	Posts in Brazilian Portuguese with translations into English	–	Posts/Videos English and Brazilian Portuguese used in separate posts. Twitter in Brazilian Portuguese with translations into English	Posts in Brazilian Portuguese with translations in English. Posts in English
	Followers	Some comments in Brazilian Portuguese. Likes and shares	Posts in Brazilian Portuguese. Videos in Brazilian Portuguese	Posts in Brazilian Portuguese	Posts in Brazilian Portuguese. Some comments in Brazilian Portuguese to posts / videos in the same language. Likes and shares	Posts/Videos in Brazilian Portuguese. Some comments in Brazilian Portuguese to posts in Brazilian Portuguese/English. Few comments in English. Likes and shares
UK	Leaders	Posts in English for events in English, in Brazilian Portuguese for events in Brazilian Portuguese. Translanguaging	Posts in Brazilian Portuguese with translations into English	Posts in Brazilian Portuguese	Posts in Brazilian Portuguese with translations in English. Twitter in Brazilian Portuguese with translations into English	Posts in Brazilian Portuguese. Posts in Brazilian Portuguese with translations in English
	Followers	Comments in Brazilian Portuguese. Likes and shares	Comments in Brazilian Portuguese. Likes and shares. Videos in Brazilian Portuguese	Comments in Brazilian Portuguese. Likes	Comments in Brazilian Portuguese. Likes and shares	Comments in Brazilian Portuguese. Likes and shares

Portuguese. The ones shared on the Italian Facebook page have subtitles in the local language. The ones shared on the American and British pages are not. The different spaces given to the Portuguese language by these churches illustrate how LPP varies according to time and space, as advocated by Blommaert (2007) in his notion of sociolinguistic scales. Portuguese is part of the linguistic repertoire of the Brazilian faith leaders and most of their followers, who are also Brazilian. Having migrated, Portuguese keeps its value for communication among the Brazilian migrants. The faith leaders in the USA and the UK seem to understand this and allow videos in Portuguese to be posted on their Facebook pages. If we consider Figure 3.2, it could be said that the LPP decisions being made in the USA and the UK in relation to the languages being displayed in their videos, are situated within the lower level scales of LPP, whereas the ones made in Italy are located within the higher level scales. The faith leaders in Italy privilege the use of the Italian language over the use of Portuguese in order to homogenise the language use of their followers in relation to the language of the wider society.

The effects of applying uniformity in LPP can be seen in the way that languages are chosen by the Facebook users to share their faith in Christ (i.e. witnessing). Italian is definitely favoured by both the leaders and their followers who participate on the Facebook page of the Italian branch. Nonetheless, the followers who comment on the posts are very few, especially in relation to the numbers on the Facebook pages of the branches in the other three countries, i.e. Portugal, the USA and the UK. Portuguese is used by both the leaders and their followers in these three countries. In other words, the different values attributed to the Italian and the Portuguese languages become very clear in relation to the posts for witnessing. Sharing one's faith in Christ is closely linked to emotions and to previous experiences. If we consider the fact that most of the churches' members are originally Brazilian, it is not surprising that they choose to communicate such personal experiences in Portuguese, the language they grew up with and that they draw upon as a resource to practise their faith. The use of Portuguese here reflects the lower scale levels of Blommaert's (2007) concept, i.e. it is linked to specific moments in their faith and represents their personal and subjective experiences in talking about them. This is true of the posts on the Portuguese, the American and the British Facebook pages, not the Italian one. Moving across to another country, the Portuguese language loses its value and is replaced by the Italian language. This change in space moves the 'order of indexicality', i.e. the norms of communication among the Brazilian Pentecostal migrants when interacting online among themselves, their leaders and possibly their new faith Italian 'siblings'.

The proclamation of Christ's good news (i.e. evangelisation) is mainly made by the local leaders and by the main pastor, who is based in Brazil, not usually their followers. The main pastor's tweets are shared on the

Facebook pages of all four international branches in question. His original message is in Portuguese and shared as such. However, translations into the local languages are provided for the posts published in Italy, the USA and the UK. A peculiarity of the British branch's Facebook page is that it is the only one that started to receive separate tweets – one in Portuguese and one in English – for the same messages directly from the main pastor in April 2015. Independently of being sent in one single or two separate posts, the translations of these messages highlight the fact that '[t]he dissemination of religious belief to new communities is a fundamentally linguistic act' (Liddicoat, 2012b: 127). The conversion of non-believers (i.e. proselytisation) is central to the activities of Pentecostal churches. Therefore, language planning in order to reach out to the local communities of the countries where the international branches are based is essential for the migrant churches. The Brazilian Pentecostal leaders, however, have not totally lost sight of the fact that most of their followers are Brazilian migrants. Consequently, the use of Portuguese is not replaced by the use of English or Italian. Instead, the languages of the migrants and the host communities are used as ways of addressing both of their target groups – speakers and non-speakers of Portuguese in Italy, the USA and the UK. This is a clear illustration of how the different sociolinguistic scales interact with each other, as advocated by Blommaert (2007).

Nonetheless, the dynamics of language use change again when the purpose of the communicative events online is to give thanks to God (i.e. praising). Praising on Facebook follows the face-to-face practice of call and response. It means that when a leader or a follower praises God, somebody responds to reinforce that praise. Both leaders in the American and British branches include English in their posts to praise God. The American branch posts in English and in Portuguese with translations into English. The British branch posts in Portuguese and in Portuguese with translations into English. However, the vast majority of their followers respond in Portuguese in both branches. These responses reflect the language negotiations which took place in the Welsh school studied by Musk (2010). In Musk's study and in this one, languages were renegotiated by individuals when they participated in meso-level language planning, i.e. public institutions such as schools and churches. Moreover, the higher level of participation of the church members in Portugal, the USA and the UK indicates that their perception of the flexible status of Portuguese contributed to them being more active in the communicative events online. In the case of the members of the branch in Italy, the level of participation of their members was very low. The fact that this branch, as described above, appears to impose LPP of higher scale levels seems to impact on how active their members are online. Perceiving Portuguese to be the language of low status in this 'Italian' online space, the members assume a passive participation in which they 'like' the posts but tend not to post their own messages.

In sum, the use of Facebook by the four Brazilian Pentecostal migrant churches illustrates (Figure 3.1) how '[r]eligious considerations can influence language choices and practices in the secular sphere, while secular issues relating to language can influence choices and practices in the religious sphere' (Liddicoat, 2012a: 75). The theological orientation of the church's four international branches has led them to make decisions about LPP for their Facebook pages in order to convert local members into their Pentecostal faith. That is to say, the local languages (English and Italian) are legitimised as accepted languages to be used online to perform religious communicative events. The fact, however, that Portuguese is assigned a lower status in relation to Italian impedes the same level of participation that the members of this branch of the church have in relation to the Facebook pages of the other branches. These other churches, although creating space for the local languages and therefore for the local members, do not devalue the importance that Portuguese has to its migrant members. The Facebook page of the church in the UK is particularly successful in actively engaging a larger number of members online. This success is due to its flexible approach to bilingualism, which is reflected in the deliberate choices it makes (and allows its members to make) about which languages to adopt in order to best experience and express their faith.

Notes

(1) See http://www.statista.com/statistics/272014/global-social-networks-ranked-by-number-of-users (accessed 2 June 2015).

(2) See http://www.socialbakers.com/blog/1064-top-10-fastest-growing-facebook-languages (accessed 2 June 2015).

(3) See http://www.publico.pt/cultura/noticia/o-portugues-conquistou-a-internet-e-agora-quer-ser-lingua-oficial-nasorganizacoes-internacionais-1606507 (accessed 2 June 2015).

(4) See http://www.statista.com/statistics/244936/number-of-facebook-users-in-brazil (accessed 2 June 2015).

(5) See http://portuguese-american-journal.com/report-portuguese-is-the-third-most-used-language-on-facebook-socialbakers/ (accessed 2 June 2015).

(6) See http://www.socialbakers.com/statistics/facebook/pages/total/brazil/community/religion/ (accessed 2 June 2015).

(7) Faith leaders here refer to the managers of the Facebook page. It is not clear whether the pastors manage the page themselves. Due to the structure of the church, however, it is certain that all the activities on the page require the approval of at least the local pastors, and possibly the main pastor in Brazil.

References

Bell, A. (1984) Language style as audience design. *Language in Society* 13 (2), 145–204.

Blackledge, A. and Creese, A. (2010) *Multilingualism*. London: Continuum.

Blommaert, J. (2007) Sociolinguistic scales. *Intercultural Pragmatics* 4 (1), 1–19.

Blommaert, J., Collins, J. and Slembrouck, S. (2005) Spaces of multilingualism. *Language & Communication* 25, 197–216.

Blommaert, J., Creve, L. and Willaert, E. (2006) On being declared illiterate: Language-ideological disqualification in Dutch classes for immigrants in Belgium. *Language & Communication* 26, 34–54.

Browne, D. and Uribe-Jongbloed, E. (2013) Introduction: Ethnic/linguistic minority media – What their history reveals, how scholars have studied them and what we might ask next. In E.H. Gruffydd Jones and E. Uribe-Jongbloed (eds) *Social Media and Minority Languages: Convergence and the Creative Industries* (pp. 1–30). Bristol: Multilingual Matters.

Cormack, M. (2013) Concluding remarks: Towards an understanding of media impact on minority language use. In E.H. Gruffydd Jones and E. Uribe-Jongbloed (eds) *Social Media and Minority Languages: Convergence and the Creative Industries* (pp. 255–265). Bristol: Multilingual Matters.

Cunliffe, D., Morris, D. and Prys, C. (2013) Investigating the differential use of Welsh in young speakers' social networks: A comparison of communication in face-to-face settings, in electronic texts and on social networking sites. In E.H. Gruffydd Jones and E. Uribe-Jongbloed (eds) *Social Media and Minority Languages: Convergence and the Creative Industries* (pp. 75–86). Bristol: Multilingual Matters.

Darquennes, J. and Vandenbussche, W. (2011) Language and religion as a sociolinguistic field of study: Some introductory notes. *Sociolinguistica* 25, 1–11.

Ferguson, J. (2010) Shäwthan Dän, Shäwthän Kwänjè: good people, good words: Creating a dän k'è speech community in an elementary school. *Current Issues in Language Planning* 11 (2), 152–172.

Gruffydd Jones, E.H. (2013) Minority language media, convergence culture and the indices of linguistic vitality. In E.H. Gruffydd Jones and E. Uribe-Jongbloed (eds) *Social Media and Minority Languages: Convergence and the Creative Industries* (pp. 58–72). Bristol: Multilingual Matters.

Hornberger, N. (1989) Continua of biliteracy. *Review of Educational Research* 59 (3), 271–296.

Hornberger, N. (2002) Multilingual language policies and the continua of biliteracy: An ecological approach. *Language Policy* 1 (1), 27–51.

Hornberger, N. and Johnson, D.C. (2007) Slicing the onion ethnographically: Layers and space in multilingual language education policy and practice. *TESOL Quarterly* 41 (3), 509–532.

Hult, F. (2010) Analysis of language policy discourses across the scales of space and time. *International Journal of the Sociology of Language* 202, 7–24.

Johnson, D.C. and Ricento, T. (2013) Conceptual and theoretical perspectives in language planning and policy: Situating the ethnography of language policy. *International Journal of the Sociology of Language* 219, 7–21.

Liddicoat, A. (2012a) Language planning and religion. *Current Issues in Language Planning* 13 (2), 73–75.

Liddicoat, A. (2012b) Language planning as an element of religious practice. *Current Issues in Language Planning* 13 (2), 121–144.

Lundby, K. (2011) Patterns of belonging in online/offline: Interfaces of religion. *Information, Communication and Society* 14 (8), 1219–1235.

Ministério das Relações Exteriores (MRE) (2012). Brasileiros no Mundo. See http://www. brasileirosnomundo.itamaraty.gov.br/a-comunidade/estimativas-populacionais-das-comunidades/APENDICE%20Diplomacia%20Consular%20-%20Brasileiros%20 no%20Mundo.pdf (accessed July 2015).

Musk, N. (2010) Bilingualisms-in-practice at the meso level: An example from a bilingual school in Wales. *International Journal of the Sociology of Language* 202, 41–62.

Ricento, T. (2000) Historical and theoretical perspectives in language policy and planning. *Journal of Sociolinguistics* 4 (2), 196–213.

Ricento, T. and Hornberger, N. (1996) Unpeeling the onion: Language planning and policy and the ELT professional. *TESOL Quarterly* 30, 401–427.

Sanchotene, S. (2011) A religião on-line na pós-modernidade. *Estudos de Religião* 25 (41), 167–184.

Seidlhofer, B. (2005) Key concepts in ELT – English as a lingua franca. *ELT Journal* 59 (4), 339–341.

Souza, A. (2014) Technology and language shift: The case of two Brazilian faith settings. In D. Mallows (ed.) *Language Issues in Migration and Integration: Perspectives from Teachers and Learners* (pp. 135–150). London: British Council ESOL Nexus.

Souza, A. (2015) Language and faith encounters: Bridging language–ethnicity and language–religion studies. *International Journal of Multilingualism* 13 (1), 134–138.

Souza, A. and Evans, Y. (2015) *Desafios no Dia-a-dia: Experiências de Brasileir@s no Reino Unido.* London: GEB/Queen Mary/UCL, Institute of Education. See https://geb2008. files.wordpress.com/2015/07/souza-evans-2015-desafios-no-dia-a-dia.pdf (access 25 October 2015).

Souza, A., Kwapong, A. and Woodham, M. (2012) Pentecostal and Catholic churches in London – The role of ideologies in the language planning of faith lessons. *Current Issues in Language Planning* 13 (2), 105–120.

Spolsky, B. (2012) What is language policy? In B. Spolsky (ed.) *The Cambridge Handbook of Language Policy* (pp. 3–15). Cambridge: Cambridge University Press.

Tagg, C. and Seargeant, P. (2013) Audience design and language choice in the construction and maintenance of translocal communities on social network sites. In C. Tagg and P. Seargeant (eds) *The Language of Social Media: Identity and Community on the Internet* (pp. 161–185). Basingstoke: Palgrave Macmillan.

Woods, A. (2006). The role of language in some ethnic churches in Melbourne. In T. Omoniyi and J. Fishman (eds) *Explorations in the Sociology of Language and Religion,* (pp. 197–212). Amsterdam/Philadelphia, PA: John Benjamins.

Zimmer, M. (2010) 'But the data is already public': On the ethics of research in Facebook. *Ethics and Information Technology* 12, 313–325.

Part 2

Faith, Language and Transnational Online Practices

4 Ifa, the Word and the Virtual World: A Study of the Perceptions of and Attitudes toward Ifa Religious Tradition on the Internet

L. Oladipo Salami

Traditional religious practices among the Yoruba, whether in supplication or consultation, are essentially oral or verbal. Although today, some written Ifa 'liturgy' or 'scripture' exists, the spoken word is by far the most important medium for rituals and supplications among devotees of Ifa. Church leaders are beginning to use the internet for evangelization, spiritual counseling and live streaming of church services, including sermons and prophetic messages. This chapter reports on an investigation of the perceptions and attitudes of Ifa devotees in Nigeria towards the use of the internet in Ifa religious worship. This is particularly against the background that, like many other religions, the essence of Ifa lies in the sacredness and verbalization of its liturgy. Thus, the overall purpose of this study is to seek to find out the likely impact of the introduction of internet platforms on Ifa religious practices.

Introduction

The Yoruba people are known to be one of the largest ethnic groups in Nigeria. There are smaller groups in countries like Benin and Togo (Hallegren, 1988), and Yoruba descendants can be found in Brazil, Cuba and Trinidad and Tobago. In their religious tradition, the Yoruba are often seen, like many other African traditional religions, to be 'this worldy' in the sense that theirs is not a religion of salvation. Thus, they acknowledge in their religious tradition that the human problems to be solved are of this world, this life and its continuity (Hallegren, 1988: 9). Ifa is considered a very important aspect of Yoruba tradition and culture, with its goal to solve problems and overcome the challenges necessary to live a good life. According to Verger (1982), Ifa is a system that exists to care for the spiritual needs of the Yoruba people in joy and in sorrow and in all practical concerns of life (Feuser, 1982: iv). The Yoruba believe that Ifa can help in solving human problems through divination and thus they seek its assistance. Hallegren (1988) notes that the followers of *Orisa* Ifa (the deity) worship the mythical founder of Ifa called *Orunmila*. Worship and Ifa divination are based on certain rituals and the power of the spoken word. The spoken word carries the prayers, invocations and chants of the diviner (*babalawo*) and of the worshippers or devotees. Ifa divination, as part of the Ifa or *Orunmila* religion, is traditionally based on verbal interactions between the *babalawo* and his 'clients' because, for the Yoruba, the spoken word carries power. In divination, this power can be seen in the throwing of the *opele* or the divination chain (much like throwing dice or shuffling Tarot cards) and interpreting its results. However, the use of information and communication technology (ICT), especially the internet, has grown tremendously across the world in recent years, trying, seemingly, to reject the spoken language as well as face-to-face interactions, and leading to the merging of religious practices. The internet is not only being used for simple information exchange, but also as a tool in social and cultural activities, including community building, cultural revival, political mobilization, social campaigns and community sensitization programmes such as the dissemination of government activities and the activities of non-governmental organizations and civil societies on social media. These may include health and sanitation and immunization programmes.

In a study of the role of computer-mediated communication (CMC) and information in religion, Hackett (2006: 67) observes that the internet has not only transformed the study of religion but also the practice of religion itself through its multifunctional roles. These roles, in the religious domain, she notes, include communicating, informing, learning, experiencing, practicing, seeking, commodifying, advocating, healing and problem-solving. Also, Campbell (2011: 2) mentions that online religion or religious practice online is taking place within networked communities, leading to the development of storied identities; causing some shift among authorities in religious institutional structure; encouraging convergence of different religious practice due to access to different interpretations and rituals on the internet even within the same religion; and leading to multisite realities arising from the existence of online and offline spaces for religious practice. She observes further that online religion represents certain cultural and social changes occurring in society, as an online religious community contains a network of social relationships, structures and patterns of belief that are highly malleable, global and interconnected. In this network, she holds, the internet has caused traditional hierarchies to become flat, encouraging instantaneous communication and response and widening access to sacred information.

In other words, it can be surmised that the internet has changed the traditional ways of interaction between and among religious practitioners. Online religious practice tends to allow for personalized rather than institutional narratives of faith. This is especially true in blogs where one can often find personal narratives or narratives about personal experiences.

Campbell and Connelly (2012: 434) note that as technology became more user-friendly to a wider audience in the 1980s, it began to have an impact on religion. They note thus that religion began to show its presence on the internet in the form of early online religious communities such as Net.religion, Ecunet.org and Buddhanet.net that featured chat rooms and discussion boards. Furthermore, it is observed that throughout the 1990s, religion expanded tremendously across the internet landscape, with web pages containing religious content increasing from 1.7 million in the late 1990s to about 51 million by 2004 (Campbell & Connelly, 2012: 434–435; citing Hojsgaard & Warburg, 2005). At this time, religious activities also started appearing on blogs and other online virtual worlds. For Campbell and Connelly (2012: 435), these varied forms of religious practices showed the growth and acceptance of the internet by many religious groups and individuals.

From the foregoing studies and reports, it is clear that religious practices have begun to transit into virtual realms across the globe. The spread of the use of CMC and information technology to religious activities has also begun to impact religious practices in Nigeria, especially within the Pentecostal Church, where leaders are beginning to adopt ICT. This includes using mobile phones and the internet for evangelization, spiritual counseling and live streaming of church services, sermons and prophetic messages. In other words, there is an emerging online Christianity in Nigeria wherein churches are embracing e-services due to increased membership in their congregations and increased physical distances to churches and places of worship (Hackett, 2009).

In the case of Yoruba traditional religion, it is observed that during the annual *Osun* and *Odun* Ifa festivals in Osogbo and Ile-Ife, respectively, devotees of Ifa travel from distant places outside Nigeria, such as Brazil, Cuba, Jamaica, Trinidad and Tobago, the USA and Western Europe (Germany) for the purpose of worship, celebration, knowledge acquisition and initiation. During their visits to Osogbo and Ile-Ife in Nigeria from these distant places, foreign devotees are likely to incur more expense in time and money than if they were to worship via internet platforms from anywhere in the world. There are therefore devotees from beyond Nigeria and outside the continent of Africa who may not be able to physically participate in religious ceremonies within these centers of worship, but by using mobile phones and the internet, it can be assumed that the potential costs incurred from traveling and physical participation in Ifa ceremonies (in Nigeria) can be minimized. This study, therefore, sets out to investigate the perceptions and attitudes of Ifa devotees in Nigeria, tapping into the 'advantages' that

CMC and information technology, particularly the internet, can confer by establishing, for example, platforms for Ifa religious practice and worship online.

Ifa Religious Tradition among the Yoruba

Ifa is a part of the larger Yoruba religious system that encompasses the propitiation of a number of divinities and/or deities including Orunmila (giver of wisdom/oracle divinity), Osun, Oya, Sango, Ogun and so on. As noted earlier, Orunmila is seen as the *orisha* or the deity of wisdom and knowledge. Thus, the worship of Orunmila is the worship of Ifa and vice versa. Aspects of Ifa religious tradition include cleansing, healing, counseling and divination. Ifa devotees hold that Ifa is a system of guidance that gives them genuine and unequivocal direction on the path to take in their lives. It is their belief that Ifa is the 'word' (*ohun*) of Olodumare (the Supreme Being or God) encompassing all knowledge of things past, present and the future. Although divination is central to Ifa religious practice, it is this specific aspect of the religion that provides guidance to other activities like initiation, cleansing, healing, offerings and so on. Since the major goal of the religion is for humans to live in harmony with themselves and the forces of nature in order to achieve their destiny (*ori*) in life, Ifa divination becomes critical to achieving this goal.

Ogunade (2010) notes that worship has become a way of life to the Yoruba people:

> [Because] every major event in their lives is attended by elaborate worship. For example in typical formal worship libation is poured to open up the earth for the spirit beings to attend to the worship; invocation of esoteric names, praises and formulae which attract the beings of worship, and Divination, by which means the message and revelations of the Being of Worship are made known to people. Worship is given its character also, in sacrifices (of animals and in the past, of human being). (Ogunade, 2010: 2)

It is important to note that Ifa religious tradition is, essentially, an oral tradition wherein speaking, chanting and invocation take place. Ifa is a tradition that involves divination that can be performed for individuals or entire communities (villages, towns and lineages). This divination can take place at either a *babalawo*'s home altar or shrine or within a community or town's shrine or temple. Divination is a critical aspect of Ifa tradition because it seeks to provide knowledge about the *ori* (destiny) or life path of the individuals or community concerned. The inquiry about the individual's path in life via Ifa is assumed to assist in guiding the individual in how to live in harmony with humanity as well as in cooperating with or propitiating what are referred to as 'the forces of the universe'. This, essentially, is the goal of the religion. As I have noted elsewhere (Salami, 2006: 101), there seems to be no meaning in life for the Yoruba without religion.

The process of divination or *ifa dida* (throwing the opele) in Ifa tradition involves a chat with a *babalawo* (Ifa priest who is usually a trained diviner) who will offer prayers and seek the permission of the *orishas* (divinities) to attend to the problems or inquiries of a client, who during the process may be asked to hold in his/her palms or to shake certain items used in divination. The reason for this is said to be that the personal spiritual energy of the client must be brought into the process of divination which is believed can be tapped into by holding or shaking an item of divination. The tools of divination, that is *ifa dida*, are opele (a string consisting of eight disks) and *opon ifa* (a wooden tray) which the *babalawo* uses. Divination may also involve the use of palm kernel nuts (called ikin) which are manipulated to provide *odu* or verses of Ifa which may contain the possible message, counsel, rituals, offerings, etc., for the client because, very often, Ifa prescribes rituals or offerings as parts of the solutions to human problems. Although during the course of the fieldwork for this study, we were informed by a *babalawo* – Chief Agboola – that there is currently a plan to compile a book about Ifa worship (featuring scripture and liturgy), it needs to be mentioned here that at the time of writing, there is no formal Ifa liturgy or a book of practice. This is not to say, however, that today there is no documentation at all of Ifa corpus or verses (*odu*). Ifa religious practices are primarily of the oral tradition, not the written word. Thus, an interaction between the diviner includes the recitation or chant of the *odu ifa* that comes as a result of throwing the *opele* or *ikin* manipulation such as in the following Odu Ifa example:

Obara Ogunda
Amodun o jinna
Ke'ni o ma hu eebu je o
Dia fun Orunmila
Ti yoo t'ori Akapo o re bo'nu igbo
Ebo ni won ni ko waa se
O gb'ebo o ru'bo
Igbati Orunmila t'ori Akapo re bo'nu igbo
L'aje waa de gbuurugbu
Ero Ipo ati Ofa
E waa b'ani ba'yo
E waa wo're

Translation
Next year is not far
Let no one eat what they are supposed to plant
Ifa's message for Orunmila
When going to subject his Akapo to Ifa initiation rites

He was advised to offer *ebo*
He complied
When Orunmila subjected his Akapo to Ifa initiation rites
The *ire* of financial success came trooping in
Travellers to Ipo and Ofa towns
Join us in the midst of joy
Come and perceive all *ire* of life
(*ebo* means 'sacrifice' and *ire* means 'good')

In the odu above, it can be read that Ifa's counsel is that there is a need for the client to plan for the future and seek the help of the divinities through offerings to achieve success. It is not unlikely that in the process of chanting this *odu*, the *babalawo* will evoke some emotion in the client while waiting for the answer to his/her request. It can thus be surmised that divination as a central aspect of the Ifa religious tradition involves physical objects, physical space and a physical presence. The question then is how would a client seeking answers to life's challenges from a *babalawo* (through *Ifa dida*) engage in this process via the internet?

In a study of the documentation and propagation of Ifa on the internet, Olajubu (2006) observes that although it may be difficult to state exactly when Ifa first appeared online, it now has a large presence. She says that the internet has impacted very much on Ifa tradition in that:

> Globalization presumes that geography is no longer a limitation for human contact and interaction. Central to this assertion is the Internet with the many provisions it encompasses. Hence, in the contemporary society, consultation on any issue of human endeavor is now possible through e-mail, telephone calls and fax messages. (Olajubu, 2006: 7)

For Olajubu, the documentation and propagation of Ifa on the internet has implications for Ifa religious tradition. Firstly, the online presence of Ifa has enabled the shrinking of space between Yoruba Ifa adherents in Nigeria and non-Yoruba adherents outside the continent of Africa. Secondly, the development has led to innovation, fusing together Western orientation and Yoruba religious traditions. Thirdly, animal sacrifice (*ebo*), which was taken for granted by Yoruba Ifa adherents, has been placed in the public domain of debate in the West by animal rights groups.

Ifa Presence on the Internet

As mentioned earlier, we conducted a web search for the presence of Ifa on the internet. Although searches online hit thousands of Ifa-related topics, it should be noted that there are very few websites devoted to online Ifa religious traditions deriving from actual Yoruba Ifa *babalawos*. There are also not many websites built, run or hosted by Nigerian-based Yoruba priests or priestesses devoted to online Ifa religious tradition. There are, however, hundreds of articles and references to the subject of Ifa in relation to African religious tradition, Yoruba culture and online Ifa by devotees outside the continent of Africa. There are also Facebook and Yahoo (discussion) groups and adverts such as www.facebook/ifa-esoteric-school and OmoOdua@yahoogroups.com, which are concerned with training and discussion about Ifa. For example, the website Spiritual Solipsism, run by Jaap Verduijn, focuses on general spirituality but features some references to Ifa religion. In this website, the owner describes how the rituals of *ibori* (feeding the head/giving energy to the head) can even be carried out alone, whether recommended or not by divination. Thus, for example, if a ritual involves the use of cool water (*omi tutu*), the prayer to be said over the head of the person who is being given energy will be

Emi ma ji loni o, o, mo f'oribale f'Olorun,
Ire gbogbo maa wa ba mi, Ori mi dami daiye,
Ngo ku mo. Ire gbogbo ni temi,
Imole ni ti Amakisi. Ashe.

Translation
Now that I am waking up, I give respect to Olorun,
Let all good things come to me, my Ori give me life,
I shall not die. Let all good things come to me.
The Spirits of Light are in the East. So be it.
(From Jaap Verduijn 'feeding the head', n.d.)

Campbell (2011) observes that one major influence of the internet on online religion is the introduction of innovation arising from experimentation, self-expression and religious lifestyles different from offline religious activities. In the invocation in the website given above, we can also see the impact of the internet on Ifa in the sense of the introduction of innovation into the Yoruba ritual. In the ritual, we see the innovative use of English to do *ibori* with Jaap Verduijn ('feeding the head') suggesting that if a non-Yoruba-speaking *babalawo* or devotee wants to do *ibori*, they should pray or invoke in the language they master best. The implication of this is to say that Ifa religious activities can be carried out in any language, not just the Yoruba language. In other words, although the Yoruba language is assumed as the language of Ifa, Ifa can 'speak' other languages. This change, arising from innovation to counter the challenge of language learning, can also be compared to, perhaps, the change from the use of the Latin language in the Roman Catholic Church in Nigeria to English. We will also observe some innovation in theological interpretations due to different perspectives that religious practitioners are able to bring to the fora provided by internet platforms where the pastor, imam or the chief priest's views may be treated as personal. Olajubu (2006) notes, for example, the influence of the internet on the place of same-sex marriage in the following extract:

> A cursory examination of Ifa texts (256 chapters) shows that no record of same sex marriages existed among the Yoruba. However, since the concept of scriptural interpretation is dynamic and dependent on the interpreter's lived experiences, same sex marriages like some other human practices which were alien to traditional Yoruba thought system have come to feature in the contemporary practice of Yoruba religion, and Ifa practice worldwide. Whether same sex marriages/relationships are sanctioned by Ifa or not would then become an issue of relativity rather than a rigid or monolithic analysis of scriptures. . . . (Olajubu, 2006: 13)

The quote above seems to also support what Campbell (2011) describes as one of the impacts the internet has had on the encouragement of experimentation and self-expression of belief and lifestyle in online religion.

From the foregoing discussion, we will assume first, that a face-to-face interaction with a *babalawo* is no longer critical because Ifa adherents or devotees can administer the ritual of *ibori* on themselves. Secondly, due to the impact of the internet, Ifa devotees may now no longer need a *babalawo* as the sole determiner of theological interpretations of Ifa, as they also now have access to varied sources in their network of Ifa practitioners online.

Divination via Mobile Phone and on the Internet

In this section, we will attempt to describe the general perceptions and attitudes towards Ifa divination on the internet that have been reported and discussed. From the website OrishaNet, it is said that

> Although there are people who advertise consultations by phone or mail, we feel strongly that the consultation will usually not have nearly the depth it would have...

The above represents the belief that the physical presence of a client for Ifa divination is best for the religion. But a priest – Ifaluade on the website www.ifaluade.com – holds that Ifa divination or consultation can be done online without physically visiting a temple or an Ifa priest. Priest Ifaluade says that all a client needs is to sign up, have an account with the website and make his/her request for Ifa divination online. The reports or responses of Ifa from the divination will also be given online, not immediately or spontaneously, but rather in a day or two. Furthermore, Priest Ifaluade says that sacrifices prescribed by Ifa via internet consultation are done on behalf of clients rather than the clients being asked to carry out the sacrifice themselves. In other words, the spoken words of Ifa in terms of invocations are being supplanted by the written or digital word. This contrasts with the practice of chanting or invocation with which the practice of Ifa divination has been known, especially among the Yoruba in West Africa. In a post on the news website www.Gbooza.com, a *babalawo* based in Lagos, Nigeria, notes that

> The procedure [in Lagos] is to visit the website and fill a consultation form online. It is stipulated that one needs not indicate his problem as the priest will use the bio data supplied to consult, identify the trouble and proffer solutions.

This is also to show that one does not need to visit or see an Ifa priest physically before one can consult. It is reported also on the Gbooza news website that in Lagos and its surrounding towns, more Ifa priests are making Ifa consultation technology driven, thereby making it accessible to many people, as demonstrated by Ajigbotifa Temple (www.Ajigbotifatemple. com), which states in their mission:

> ... to make man live a fulfilled life. To professionally help man with the help of Olodumare and Ifa; to help man in overcoming both physical and spiritual problems, give foresight into the future, prosperity, health, good children, peace, love and more by communicating with us through the Holy Odus of Ifa...

It is also pointed out that apart from consultation or divination over challenges (physical and spiritual), the Ajigbotifa Temple carries out initiations as well as online worship.

In her study, Olajubu (2006) observes that on the internet,

> Fake products have been known to circulate in the name of authentic spiritual items, just as psychology is sometimes substituted for Ifa divination through e-mail, telephone calls or fax messages. (Olajubu, 2006)

Apart from this type of report given by Olajubu, one still holds that the essence of Ifa is contained in the believed sacredness of the use of language or the verbalization of the *odu ifa* (Ifa verses) by *babalawos*, which may be lost in an internet practicing platform where technology filters the spoken words of the *babalawo*. However, Ifa religious tradition has fewer places of worship and fewer devotees among the Yoruba of southwestern Nigeria than the Christian or Islamic religions. Ifa also faces the challenge of disclosure, especially because of the stigma attached to worshipping and consulting an *orisa* or a traditional divinity or deity. This is perhaps why not as many people as would probably like to participate actively in Ifa religious tradition do so; because they would not want to be stigmatized. It may, therefore, be a plus for Ifa to be available on the internet for those who would practice in the privacy of their homes.

The present study

In examining the perceptions of and attitudes to virtual or online Ifa, we were guided by the following questions:

(1) Where do *babalawos* and Ifa devotees worship?
(2) How regular is their worship routine?
(3) How far, through a website survey, has CMC and information technology penetrated Ifa religious tradition among the Yoruba? In other words, how far along is the transition from 'brick and mortar' to virtual places of worship?
(4) How do both the devotees and non-devotees of the Ifa religious tradition perceive the use of CMC and technology in the Ifa tradition? Knowing that divination is a central aspect of Ifa religious tradition, how do *babalawos* as well as non-adherents perceive Ifa practice via technology? That is, being a largely oral religious tradition, what is the level of acceptance of technology in Ifa religious practice?

The above questions were at the background of this study to assist us in establishing the extent of the practice of Ifa in the location of the study as well as investigating its potential transformation to the virtual space.

Data Collection

The methodology for the present study involved a web search and the use of a questionnaire survey with close- and open-ended questions. The survey was supplemented by in-depth interviews with a select number of Ifa priests, *babalawos*. The web search was to investigate the extent of the penetration of the internet in Ifa religious tradition. The sample population for the interviews was made up of 25 *babalawos* and 20 non-devotees of Ifa from the Yoruba-speaking western part of Nigeria. It is perhaps pertinent to mention that the *babalawo* priesthood in Yorubaland is not an established or public institution like the Christian priesthood and, as such, does not have any general or specific marker(s) of identity except, perhaps, personal disclosure. In this investigation, we used the friend-of-a-friend (network) approach to select the priests interviewed. This was done because many of those we encountered in the communities we visited denied knowing any *babalawo* when asked, and many of them were not ready to participate in the study as informants or interviewees on the subject of the Ifa religious tradition. It may be worth mentioning that gathering data on Ifa among the generality of the Yoruba people in Nigeria is seemingly a difficult enterprise. This is particularly so because asking about a *babalawo* in the community is like asking whether one knows, in the perception of the largely Christianized and Islamized Yoruba public today, some 'bad' individuals. This perception grew out of the encounter of Yoruba traditional religious practices with both Christianity and Islam at their coming. They had both, to a large extent, portrayed these practices as not so godly or outrightly pagan, and therefore, as not to be engaged in. For representativeness of Ifa practice across Yorubaland, we visited the five states of Ondo, Ekiti, Oyo, Ogun and Osun in the southwest region of the country to sample five *babalawos* from each, totaling, as mentioned earlier, 25 *babalawos*. We interacted with them, seeking to learn about their training as Ifa priests; their level of literacy (especially computer/internet literacy as this may have influenced their attitudes); and Ifa worship, divination and other aspects of the religious tradition such as sacrificial offerings and initiation rituals. We also interviewed the 20 non-devotees on what they knew of Ifa, and if they would like to consult Ifa online if available and why?

In summary, the aim of the investigation was to explore, through the survey, interviews and interactions, the general impact of CMC and information technology on Ifa religious tradition in Yorubaland, and how much the Yoruba *babalawo*, the Ifa tradition as well as non-adherents of Ifa have been influenced.

Ifa Websites

Olajubu (2006) was able to source only 15 websites that provided information on the Ifa corpus and some other aspects of Yoruba religious tradition relating to Ifa. However, in our internet search of June 2015, we found 38 websites devoted primarily to Ifa. These sites can be categorized into eight subheadings by what they claim to do: Training; Worship; Divination/ Counseling; Festivals/Ceremonies; History/Information; Chat or Discussion Groups; Initiation; and Online Store/Tools market. It should be noted that some of these sites perform multiple functions, as a site could serve more than one purpose. For example, a site could do both training and information sharing or worship and divination. The 38 websites are categorized in Table 4.1.

Our examination of the websites shows that a preponderant majority of them are hosted by *babalawos* from outside Nigeria. These are *babalawos* based mainly in the United States, Germany, Cuba and Brazil. Only nine, or 34%, of the 38 websites found are hosted by Nigerian-based Yoruba *babalawos*. In other words, for Nigerian-based Yoruba *babalawo*, the internet is yet to become a critical medium of use. However, later in this study,

Table 4.1 Websites devoted to Ifa

Focus	Number of sites
Training	13
Worship	9
Divination/counseling	17
Festivals/ceremonies	6
History/information	13
Chat/discussion groups	4
Initiation	6
Online store/tools market	6
Live streams	–

we tried to find out if the level of literacy and competence in computer/ internet usage of the Nigerian Yoruba *babalawos* we interviewed could be a factor in their inability to set up websites for Ifa activities, including Ifa worship and divination.

Place, Regularity of Ifa Worship and Divination

In the five states of the region occupied by the Yoruba people in Nigeria, the Ifa priests or *babalawo* interviewed for the present study reported here provided the information given in Table 4.2 with regard to the regularity of *Ifa bibo*, that is, the propitiation or worship of Ifa.

The daily propitiation takes place in the shrines of the individual devotees while the weekly and yearly worships take place in the different temples found across the region.

Interactions with the *babalawos* show that all of them worship at home in their personal shrines or 'altars' and only go to community shrines or temples every five days. Today, annual worship ceremonies or festivals are held by all *babalawos* in community or township temples, which are relatively recent

Table 4.2 Regularity of *Ifa Bibo* (worship/propitiation)

State	Daily	Weekly	Both
Ekiti	3	1	1
Ogun	2	3	–
Ondo	2	1	2
Osun	3	1	1
Oyo	2	3	–
Total	12	7	6

Figure 4.1 A *babalawo's* shrine

developments. The world temple of Ifa, to which all devotees from across the world come annually, is at Oketase in the city of Ile-Ife, Nigeria. All the *babalawos* we interacted with for this study claim to worship in their homes everyday while they attend group or congregational worship occasionally (Figure 4.1). It can be inferred that this might be because weekly worship is an emerging practice in Ifa tradition patterned after Christian and Muslim's weekly congregations.

We also sought to know from the *babalawos* the frequency with which clients are physically present to consult them on Ifa divination. From their responses, we observed that they all carry out divination daily except on Ifa congregation day. This question was asked in order to see if the use of CMC and information technology had taken over some of their divination routines. All 25 *babalawos* claim that they also consult for divination requests over the mobile phone. Asked further how this is done in the absence of the client, a *babalawo* from Osun State said:

> Rather than listening to the client directly, you put the opele (divination beads) by the phone with the speaker on, so the client mentions the problem to the opele and not me after which I take over... (No 2)

It is believed that a client speaking to the *opele* conveys some energy from his/her spoken words to the *opele*, thereby energizing it to be able to uncover the secrets behind the current challenge(s) confronting the client (Figure 4.2). We therefore considered that if this is so, should *babalawos* consult through any other medium? What we observed from their responses is that

Figure 4.2 Babalawo and divination tools

Table 4.3 Computer literacy of the *babalawos*

	No	%
Literate	17	68
Non-literate	8	32
Total	25	100

they all used the mobile phone. The explanations to their responses vary from (1) giving the results of the divination; (2) using the phone to give instructions on what a client should do following divination; and (3) using the phone to pray over sacrificial offerings and so on. We also decided to ask our interviewees whether or not divination should be carried out via the internet (see Table 4.3).

It is important to note that the understanding and interpretation of online religious activities may be determined by one's knowledge and apprehension of what CMC and information technology can do. Thus, for Yoruba *babalawos* and those who share their world, some literacy or competence in the use of communication and information technology-related devices is likely to influence their perception and attitude towards internet usage in Ifa tradition. Of the 25 *babalawos* interviewed, only 8 (32%) claimed to be computer literate. Table 4.3 shows the distribution of the *babalawos* according to their computer literacy levels.

Table 4.4 Do you support divination via the internet?

	No.	%
Yes	25	100
No	0	0
Total	25	100

From Table 4.4, all 25 (100%) *babalawos* we interviewed are of the view that divination online can be effective. Some of the *babalawos* interviewed, however, see the use of the mobile phone as an interruption to the possibility of the *babalawo* uncovering the underlying psychological problems a client may have, especially as the consultation is not face to face with the *babalawo*. This same view was expressed in relation to divination on the internet:

It is not proper. It is like a pupil asking a teacher to teach from home. It is better they (clients) travel down. I don't do consultation over the phone…only notify me via calls and travel down. (No 17)

However, some of the *babalawos* also held that divination over the mobile phone or via the internet was not improper because the natural elements of earth, air and water still serve as points of contact for a *babalawo* and a client even when the computer or the mobile phone mediates their interactions:

Ifa is the babalawo esoteric words of Olodumare, so, the words are in all places with the belief of the clients, it will solve his/her problems. (No 4)
The wind reaches everywhere… an element of Ifa is the wind. (No 16)

In other words, divining, which involves chanting and invocation, is carried out on the same earth that a client shares with the *babalawo*, whether or not he/she is present at the *babalawo*'s place. The chant or invocation is also transmitted through the same air which the client breathes anywhere he/she is in the world.

A number of the *babalawos*, however, also stressed the element of faith, a major ingredient of religion, in the effectiveness of online divination:

Faith on both parts will enhance success. (No 9)
As long as there is a mutual feature (e.g. land and water). Appealing to nature goes a long way. It could be more effective with the enquirer's faith at work even more than the enquirers that come in person. (No 14)
That is God's work and work and no man understands it. (No 22)

Although the *babalawos* here may have some doubt as to the effectiveness of divination through CMC and information technology, they are of the perception that faith and God play an important role in successful Ifa divination.

Attitudes of Babalawos to Ifa Live-Streaming

Technology has afforded the possibility of broadcasting live activities on the internet which individuals can view on their personal computers. This internet broadcasting, called live-streaming, enables individuals to participate in broadcast activities in their own homes or at venues they deem convenient. Thus, for example, for Ifa worship ceremonies a devotee does not have to be present in the temple to be able to worship but can do so by simply logging on to a temple site during a scheduled worship period and join, virtually, those who are physically present. We therefore sought to know the views of Ifa devotees on the acceptability of live-streaming of Ifa ceremonies. Table 4.5 contains the results of their responses.

Although the number of *babalawos* we interviewed who are computer or internet literate is low (8 of the 25, or 32%), all of them were well disposed to the idea of live-streaming of Ifa religious activities.

Table 4.5 shows that the 25 *babalawos* interviewed supported the streaming online of Ifa worship, as it can also be broadcast via the television. A number of them noted, though, that in earlier times such a move would have been opposed by *babalawos*. In asking them for their reasons for supporting online streaming, they observed the following:

There is no secrecy because Ifa is light. (*Babalawo* No. 9)

It used to be hidden but no more. We want to spread it. (No 19)

Otun Ifa cannot be seen by all... only Osi Ifa. (No 12)
(Otun and Osi refer to 'right' and 'left')

Our forefathers prevented it (Ifa) from being known to outsiders due to ignorance but now it is made open. (No 23)

From the responses above, it is clear that the general belief among the Yoruba *babalawos*, today, is that the advent of technology can impact positively on the practice of Ifa as it has provided access for non-practitioners to learn more about and to practice Ifa. The *babalawos* also believe that technology

Table 4.5 Acceptability of streaming on the internet

	No	%
Yes	25	100
No	0	0
Total	25	100

will help in the propagation of the religion. However, just like in many other religions where the 'arcane or mystical' aspects are thought to not be open to everyone, it is still held by one of the *babalawos* in our study that not all about the Ifa religion can be seen or known by outsiders even when its ceremonies are streamed live.

Non-Adherents and Online Ifa Religious Tradition

As mentioned earlier, this sample population was made up of those who have access to the internet. Of the 20 respondents, 17 (85%) claimed they knew Ifa as a religious tradition, one claimed not to know what it was, one claimed that Ifa was something ungodly while another one said that Ifa was about divination. In other words, when compared with the negative responses we got from interacting with people in the field in Abeokuta town when we sought to ask about where we could find *babalawos* to interview, these respondents seemed to show more positive attitude towards Ifa. When we asked if they would consult Ifa for divination if available online, 7 (35%) said they would while 13 (65%) said they would not. From our interactions with the respondents, some gave the following reasons for their responses:

> The internet will be more (sic) faster (No 1)
> For the sake of privacy (No 2)
> It helps to maintain anonymity and confidentiality (No 3)

Those who claimed that they would not do internet divination gave the following reasons:

> The consultation may become exposed to other users. Hackers abound. (No 5)
> I believe in human being. It is human being that installed on the internet and there is the proverb that says the devil you know is better than the angel you don't know. (No 8)

Although it seems from the foregoing that the number of non-adherents who are well disposed to internet divination is fewer than those who are not, the reasons given by both groups are quite interesting. While those who prefer the internet consider that it has the advantage of confidentiality, those who do not want it think the *babalawo* – the human agent – is seemingly more reliable.

Conclusion

In this study, we have examined Ifa religious tradition, its presence on the internet and the perceptions and attitudes of its practitioners as well as its non-adherents to its potential as a virtual religious tradition among the Yoruba people of southwest Nigeria. The study shows three major outcomes. First, most of the current websites on Ifa religious tradition are hosted by non-Yoruba adherents of Ifa. Although we observe that some of the web hosts bear Yoruba names, these hosts are either African-Americans, Brazilians, Cubans or Venezuelans while some others are Ifa or Yoruba *orisha* devotees who have adopted Yoruba names. It is also shown, however, that the development has given rise to innovation in the tradition as we can see from the fusion of modern technology and Yoruba tradition.

Secondly, from their responses, most of the *babalawos* claim that they can attend to clients via the mobile phone or the internet without their clients' physical presence. The implication of providing consultation and services to clients without being physically present in face-to-face encounters with *babalawo* is the possibility of the reduction in the cost of services. In other words, this study confirms the observation by Hackett (2006) that the internet can help to transcend the limitations of finance and space in religious practices. However, as the *babalawos* in this study also rightly observed, the virtual presence and practice cannot substitute for the human angle of the *babalawo* in terms of what Hackett (2006, citing Goethals, 2003) describes as the expressive and emotional power of live practices. However, even though the *babalawos* in this study have mentioned that the power of Ifa is to be found in the use of the word (invocation), we can surmise here that they are at ease with the seeming complementary roles of the mobile phone divination requests as well as live streaming to the physical presence of clients.

The third issue of interest the study has shown is that from their responses the *babalawos* seem to suggest that CMC and information technology will make the Ifa tradition more open, as other people will now have access to what was previously considered its secret practices. However, the implication for language and communication in the tradition is that when the computer becomes the mediator, verbal interaction in the form of *odu Ifa* chants and the question and answer session between a *babalawo* and a client may become infrequent.

References

Campbell, H.A. (2011) Understanding the relationship between religion online and offline in a networked society. *Journal of the American Academy of Religion* 1–30.

Campbell, H.A. and Connelly, L. (2012) Cyber behavior and religious practice on the internet. In Z. Yan (ed.) *Encyclopedia of Cyber Behaviour* (pp. 434–447). Hershey, PA: IGI Global.

Feuser, W. (1982) The tales from Brazil: An introduction. In P. Verger (ed.) *DILOGUN: Brazilian Tales of Yoruba Divination Discovered in Bahia*. Translated, edited and annotated by W. Feuser and J.M. Carneiro da Cuba. Lagos: Centre for Black and African Arts and Civilization.

Goethals, G. (2003) Myth and ritual in cyberspace. In J. Mitchel and S. Marriage (eds) *Mediating Religion: Studies in Media, Religion and Culture* (pp. 257–269). Edinburgh: T & T Clark.

Hackett, J.I.J. (2006) Religion and the internet. *Communication Research Trends* 25 (1), 3–24.

Hackett, J.I.J. (2009) The new virtual (inter) face of African Pentecostalism. *Society* 46 (6), 496–503.

Hallegren, R. (1988) *The Good Things of Life: A Study of the Traditional Religious Culture of the Yoruba people*. Loberod: Plus Ultra.

Hojsgard, M. and Warburg, M. (2005) *Religion in Cyberspace*. London: Routledge.

Ogunade, R. (2010) Yoruba Religious Worship in Modern Context. See StudyLib.net/doc/197066 (accessed 18 August 2017).

Olajubu, O. (2006) The documentation and propagation of Ifa corpus and the challenge of modern information technology: The Internet as a Focus. *ORITA-Ibadan Journal of Religious Studies*, 38 (1&2), 153–165.

Salami, O. (2006) Creating God in our image: The attributes of God in the Yoruba socio-cultural environment. In T. Omoniyi and J.A. Fishman (eds) *Explorations in the Sociology of Language and Religion* (pp. 97–118). Amsterdam/Philadelphia, PA: John Benjamins Publishing Company.

Verduijn, J. Ibori: Feeding the head. See: www.angelfire.com/nv2/solipsist/practices/theifaconnection/ritualsandsuch/01c2ea92cb0ef3a05.html (accessed February 2017).

Verger, P. (1982) *DILOGUN: Brazilian Tales of Yoruba Divination Discovered in Bahia*. Translated, edited and annotated by Willfried Feuser and Jose Marianno Carneiro da Cuba. Lagos: Centre for Black and African Arts and Civilization.

Appendix

Some IFA websites:
(1) www.theyorubareligiousconcepts.com
(2) www.agolaroye.com
(3) www.ileorunmilashrineinc.org
(4)   www.ileabomale.com
(5) www.spiritualtools.org
(6) www.ileifa.org
(7) www.awonifa.com
(8) www.ajigbotifatemple.com
(9) www.egbeifaorun.com
(10) www.oyekuofun.org
(11) www.ileiwosanorunmilamimotemple.com

Websites hosted by Nigerians:
(1) www.templeifaorisa.org
(2) www.irunmole.org
(3) www.aresafaithweb.com
(4) www.fakayodefayemifatunde.wikifoundry.com
(5) www.ileifaagbaaye.com
(6) www.oracleifa.com
(7) www.ileorunmilatemple.webs.com
(8) www.ifaagbaye.org
(9) www.oyekuofun.org

5 Globalising Yoruba Taboos and their Sociocultural and Religious Values

Iyabode Deborah Akande

This chapter focuses on the social cultural values of Yoruba taboos among the Yoruba of the southwestern part of Nigeria and provides a broader classification of these taboos. Data were gathered mainly through interviews with 24 elders who volunteered to serve as informants for the research. The informants were from six different Yoruba communities, namely, Owo in Ondo state, Igede-Ekiti in Ekiti state and Ara, Ede, Oba-Ile and Ile-Ife in Osun state. These communities were purposely selected based on the accessibility and the willingness of the interviewees to participate in the research. To complement the interviews, data were also drawn from postings on taboos on Nairaland, Facebook and online news. The chapter found that taboos among the Yoruba performed certain sociocultural functions such as maintaining harmony and peaceful coexistence in society. It was also revealed that, although the categories are fluid and fuzzy, Yoruba taboos can be grouped into myth-related taboos, lineage taboos, religion-related taboos and instructive taboos based on historical, religious and functional contexts. Finally, the chapter emphasised that Yoruba taboos are no longer purely oral in mode as they have been digitalised and, as such, their discussions on digital space are characterised by computer-mediated communication (CMC) linguistic features such as phonetic spelling, code-mixing, and non-use or haphazard use of punctuation marks. The chapter concluded that many of these taboos are being translocalised since the majority of them now transcend the traditional oral space and occupy the digital technological space. The current advancement in digital and media space has now made it possible for Yoruba traditional orature, of which Yoruba taboos constitute a part, to be accessed by people from different parts of the globe.

Introduction

Taboo is one of the cultural resources that characterises almost every society. Taboos exist for every segment of human beings in society, and as there are taboos for older people, so also there are taboos for children (Adebileje, 2012; Odejobi, 2013). Taboos for children are usually didactic in nature as they teach children certain things such as how to behave properly in society. Odejobi (2013: 221) supports this by remarking that taboos can control the moral order within a society and that they originated 'from the fact that people discerned that there were certain things which were morally approved or disapproved by the deity' and that taboos are 'not contained in any written law but are preserved in the tradition'.

Most scholars who have worked on taboos unanimously agree that there is no society that does not have its own culture, and by implication, its own taboos (see Durkheim, 1963; Holden, 2000). In any society, taboos constitute one of the guiding principles that, to a large extent, control our behaviour and spell out the forbidden acts of any society. Taboos are generally observed not just to reshape our behavioural patterns, but also to engender social cohesion and peaceful coexistence in society (Afe, 2013; Jegede, 2002; Onadeko, 2008). Although every society has its own taboos, the belief in taboos and the impact they have on individuals as well as their roles in society seem to be appreciated more in rural settings and less-developed communities than in urban and well-developed societies. Almost every aspect of life has some taboos related to it. There are taboos related to gods and religion (Alade, 2007; Idowu, 1962), there are some that have to do with food (see Olarinoye et al., 2014) and there are those that deal with our environment and how to make it clean (Olabode & Siyanbola, 2013). There are also taboos on how individuals are supposed to relate to one another in society.

Yoruba taboos are found in different cultural resources. Most oríkì, forms of Yoruba panegyric poetry, contain myth-related taboos (Oluwole, 1992), while taboos are also very common in Yoruba proverbs (Aboluwodi, 2014; Akanbi, 2015). For instance, a Yoruba proverb in which there is a taboo is 'A kìí tojú ẹlẹ̀sẹ̀ mesan kàá' ('you do not tell a physically deformed person that he is deformed in his/her presence'). The Yoruba believe that if a person is physically challenged, using this disability as a way of addressing them can lead to a quarrel or even chaos in society. Thus, it is forbidden to do so as the proverb above shows. And, as will be argued, just as other scholars have, these taboos perform certain social functions without which society may not be at peace. Apart from engendering peace and harmony in society, Yoruba taboos greatly help in the upbringing and socialisation of children (Odejobi, 2013). A good knowledge of taboos is very crucial for any child who wishes to know his or her culture and how that culture can be explored for human and societal development (Ibagere, 2014).

There is a close link between language and religion and this interrelationship is evident in several studies that have been carried out in this area (see Bell, 1987; Carpenter, 1992; Wallace, 1966; Wheelock, 1982). Religious beliefs and practices are expressed through the instrumentality of language. The observance of taboos is cultural and it forms a crucial aspect of the religious belief system of the Yoruba; religious beliefs can only be explained and understood within the context of language – whether spoken, written or digital. As Crystal (1965) has observed, language plays crucial roles in the understanding and presentation of religious beliefs. Religious language is perceived to be different from everyday speech and the differences are as a result of some semiotic and pragmatic questions that users of religious language face (Keane, 1997). Keane (1997: 48) remarks that '[R]eligious observance tends to demand highly marked and self-conscious uses of linguistic resources'.

Among members of some Yoruba cults such as the Ogboni cult or the Ifa cult, certain codes and expressions are sometimes used and these codes are rarely understood even by other speakers of Yoruba who are not initiated. In other words, certain traditional religions are characterised by an esoteric use of language. Yoruba culture, of which taboos constitute an important segment, is passed down from one generation to several other generations through language; so, without language the cultural transmission of oral genres may be impossible. However, as pointed out above, the language that is considered instrumental to the spread of Yoruba religion and culture no longer needs to be oral or written; it can be digital.

Both Sawin and Souza (this volume) have examined the interconnectedness that exists between language, religion and digital media. Sawin has discussed how social media can be used by Christians to demonstrate their identities in virtual communities. He notes the linguistic practices that minority Christians engage in when communicating on virtual spaces and how such practices differ from those they use in physical church domains. Souza also demonstrates how digital technology, specifically Facebook, can be used by migrant churches as a medium for language planning. By focusing on the language choices of faith leaders while posting on Facebook and comparing the choices with those made by their followers, she argues that for migrant churches to be successful in engaging their migrant members online, they need to adopt a flexible approach to language planning.

The nexus between orature and digital technology is one that calls for serious attention as digital spaces now successfully accommodate a robust display of oral genres such as proverbs, poetry and taboos. Many oral forms are now found and properly documented on websites and are discussed on online forums; more and more spaces are being continually created by digital devices for the investigation of old oral genres. For instance, 50 years ago it would have been very difficult to imagine that various forms of Yoruba orature could appear on Facebook, Twitter and online forums such as Nairaland. The advent of hybrid digital forms and social media like wikis, blogs and databases has meant that the meanings of traditional oral genres are now digitally mediated and they can no longer be considered as purely spoken or written to be spoken (Mills, 2010). Androutsopoulos (2011a) remarks that language innovation and change in CMC is characterised by a perfect blend of both spoken and written features, the use of strategies of economy as evident in the use of acronyms and emoticons and lastly by the deployment of compensatory means for prosodic and visual clues. He argues that 'digital media enables an expansion of vernacular writing into new domains of practice, and therefore a diversification of writing styles and pluralization of written language norms' (Androutsopoulos, 2011a: 146).

Within digital spaces, it is possible for new forms of varieties of English to emerge and when they do, such language variation is technologically driven (see Rosowsky's introductory chapter in this volume). For instance, Akande and Akinwale (2010) examine the linguistic strategies that a group of students in Obafemi Awolowo University use in the composition of text messages. The work reveals that because of the restriction placed on them in terms of the number of characters, texters often use linguistic strategies such as clipping, abbreviation, initialisation, electronic or phonetic spelling and acronymy. Facebook provides an enabling digital space that can make people, especially young people, interact online and make connections with other people outside of their physical domain (Androutsopoulos, 2011a). Terantino and Graf (2011: 44) have suggested that Facebook can be integrated 'into foreign language courses to encourage target language production' since it is obvious that the internet and social media sites play crucial roles in students' lives (Terantino & Graf, 2011).

The interpretation and representation of ideas in social contexts are becoming intensively digitalised (Mills, 2010); an understanding of the messages and words in digital mode sometimes demands the understanding of emoticons and the visual, audio and gestural modes with which they appear. This suggests rightly, as Lemke (1998) has put it, that meanings of words on multimedia and digital spaces are not only included in the images used but they can be modified by those images. Arising from this is the need for current research on orature to pay attention to the nexus between orature and multimedia technologies (Liman, 2010). Given the fact that digital technology, a hybrid culture that comprises both oracy and literacy, continues to threaten the survival of traditional oral genres in modern societies, orature indeed needs to be contextualised by examining it within the digital space.

In this age of digital technology, different aspects of Yoruba culture can now be accessed on the internet as the traditional public space is gradually giving way to the contemporary media space as a result of modernisation. So, for most people, the web is a virtual stage that enables them to explore and discuss Yoruba taboos. They also sometimes engage in a metalinguistic discussion of these taboos on Facebook, Twitter and other social media.

A good example of a forum where Yoruba culture is discussed is Nairaland. Nairaland is a discussion forum mainly frequented by Nigerians at home and, transnationally, abroad. Other non-Nigerian participants on the forum are often friends of Nigerians and people who have some link with Nigeria. We therefore find African Americans, Britons, Asians and Jamaicans (and a host of others) contributing to a particular topic of discussion. On this forum, many issues of local and national relevance are often discussed. The recurrent topics of discussion on the forum usually revolve around local politics, music, Nigeria's cultures, fashion, economy and the belief systems of different ethnic groups in Nigeria. Aspects of the culture and belief systems on the forum include traditional marriage, (African) proverbs and their meanings, superstitions and taboos, the focus of this work. When a participant initiates a topic (usually called a thread) on the forum, participants from different parts of the world will start responding to the posts in the form of a discussion. Sometimes, other participants ask questions for clarification if they are not familiar with the thread. Discussions on a topic may go on for several weeks and there is no limitation on the length of a thread. Through this forum, many non-Nigerians discover a lot about Nigeria's cultures, music, language(s) and traditional religions (see Heyd & Mair, 2014).

Apart from the discussions on Yoruba taboos on Nairaland, the forum also helps as a depository of Yoruba taboos and other aspects of Yoruba culture. It is important to note here that most of the taboos on the forum can be regarded as general taboos. In the context of this chapter, general taboos are taboos that are neither city specific nor religion specific. They are taboos found across Yoruba land and are observed by the Yoruba generally. An example of a general taboo is one that forbids a Yoruba king from seeing a corpse. It is also generally forbidden for a king to cry in public even when somebody close to him passes away. Participants on Nairaland often discuss general taboos probably because they are usually more interested in the generality of Yoruba culture and less so in the cultures of sub-ethnic groups within the Yoruba as a race.

Given that background, this chapter is aimed at examining the social cultural values of Yoruba taboos among the Yoruba of the southwestern part of Nigeria as well as providing a broader classification of these taboos. In this chapter, some explanations and descriptions of how certain Yoruba taboos came into being are provided.

Review of Literature

Taboos have been viewed differently by many scholars. Wardhaugh (2006) notes that taboos are concerned with the avoidance of behaviour considered harmful to members of any society, while Ogunyemi (2007) points out that failure to comply with the taboos of a society can lead to severe punishment. According to Osei (2006), taboos are principles that regulate and direct the behaviour of individuals and the community as to how they can relate with deities and ancestors in Africa. Osei remarks that there are two senses of taboos. The first sense, which is narrow, is to see taboos from a purely cultic and religious perspective while the second sense, which is broader, is to look at taboos in their socio-economic and political contexts. However, the two approaches are two sides of the same coin.

According to Thorpe (1972), as cited in Odejobi (2013: 224), there are reasons for the development of taboos. Some of the reasons as highlighted by Thorpe (1972) are

(a)   to avoid accident;
(b)   to have respect for religion;
(c)   to have respect for elders;
(d)   to teach moral values;
(e)   to explain things that are difficult to understand; and
(f)   to obey rules of cleanliness.

Thody (1997) remarks that taboos, which are forbidden acts in society, can come under five categories: actions that people are not supposed to engage in; food that certain people are not supposed to eat; words and topics that people should not say or discuss; ideas and pictures that people are not supposed to think about or paint; and signs of what people should not look like.

Adebileje (2012) focused on the sociocultural description of some Yoruba taboos and examined the attitudes of some young people toward taboos. Adebileje (2012: 94) claimed that taboos are 'one way in which the Yoruba society expresses its disapproval of certain kinds of behaviour believed to be harmful to its members, either for supernatural reasons or because such behaviour violates a moral code'. The data used were gathered from 300 adults whose ages ranged from 45 to 70, and 300 undergraduates randomly selected from two universities: Redeemer's University in Ogun State of Nigeria and Obafemi Awolowo University in Osun State in Nigeria. The students whose ages ranged between 16 and 35 were from different departments and disciplines of the universities. The author gave them 10 carefully selected Yoruba taboos that cut across cooking, royalty, death and birth, and solicited their responses on them. Using Vygotskian's sociocultural approach as the theoretical framework, the research provided a systematic description of the selected taboos. The findings of the paper revealed *inter alia* that the lukewarm attitude of young people to taboos is a product of their parents' negligence. It was also found out that the relevance of Yoruba taboos is fading due to the influence of Christianity, Western culture and technology. The paper concluded that in order to uphold Yoruba cultural values, children must be exposed to Yoruba taboos; they must know their meanings and also the rationale for each of these taboos.

Olabode and Siyanbola (2013) investigated the use of Yoruba proverbs and taboos and how they could be used to solve numerous environmental problems confronting Nigeria as a country. By making use of sanitation-related Yoruba proverbs and taboos, the paper demonstrated how Yoruba proverbs and taboos could be used to solve Nigeria's environmental problems. While some taboos are universal, some are culture specific (Odejobi, 2013). Ibagere (2014: 24) remarked that 'taboos have their significance depending on the degree of the consequences on those who commit such taboos. A taboo in one society may have no significance in another society'. Ibagere (2014) examined how taboos and injunctions were employed to achieve social cohesion and engender harmony and development in society. The paper argued that if taboos and injunctions are properly harnessed and if the media play the crucial roles that are expected of them, corruption would reduce in society. It concluded that legislation can be used to back the revitalisation of taboos towards societal development.

In her research on taboos and superstitions among the Yoruba, Odejobi (2013) observed that all of Thorpe's reasons listed above could also fit perfectly into the categorisation of taboos in Yoruba land. She also identified, among others, Yoruba taboos to avoid accidents and wastefulness and those taboos that are observed to teach moral values and to have respect for elders and religions. The research concluded that Yoruba taboos can facilitate the development of Yoruba society as these taboos are not 'a means of creating fear into the children' but rather they 'played important roles in the traditional African Yoruba society' (Odejobi, 2013: 226).

Taboos play very crucial roles in the life of any community and because of these roles, the breaking of a taboo is considered offensive (see Madu, 2002; Ogunyemi, 2007). As a matter of fact, the breaking of a taboo is seen as an affront to the gods of the land and to the society from which such a taboo emanates (Osei, 2006). Though the classifications of taboos based on their functions are valid, the present chapter has attempted a broader classification of taboos based on historical, linguistic, religious and cultural contexts.

Taboos and Cultural Globalisation

In this era of digital technology, the whole world has become a global village (McLuhan, 1962). According to Tomlinson (1999: 2), globalisation refers to 'rapidly developing and ever-densening network of interconnections and interdependences that characterize modern social life'. Appadurai (1990) examined globalisation in terms of five scapes: ethnoscapes, financescapes, mediascapes, ideoscapes and technoscapes. Of these five, the one that is of direct relevance to this work is technoscape, which according to him, refers to the configuration and fluid nature of technology and how this configuration has made it easy for communication to reach different places in the world. For instance, through digital technology such as Twitter,

Facebook and discussion forums, ideas that were hitherto restricted to some geographical location can be deterritorialised and find themselves in other parts of the world (Akande, 2012a).

There are both linguistic and cultural flows from one continent to another: in most cases, the flows may be from the centre to the periphery as manifested by the adoption of African American Vernacular English (AAVE) by many Nigerian hip-hop artistes as well as the similarities in the way that these artistes and many other young Nigerians dress (see Akande 2012a, 2012b, 2014) and, at times, the flows may be from the periphery to the centre. Akande (2012a: 30) remarked that globalisation 'is not a one-way traffic phenomenon' as there is evidence that some words in Nigerian languages do find their way into English in other continents.

Akande (2012a) discussed different ways in which African culture (specifically Yoruba culture) has traversed beyond Africa. He pointed out that through the establishment of institutes of African languages and cultures or related institutes across the globe, African cultures have moved beyond the border of Africa. The aspect of cultural globalisation with which this chapter is concerned is Yoruba taboos. The major argument here is that since these taboos are found on discussion forums such as Nairaland and Facebook, they occupy a deterritorialised digital space that anybody from any part of the world can access. However, rather than undergoing the process of globalisation, Yoruba taboos have undergone (and are still undergoing) the process of translocalisation (see Pennycook [2007] on translocalisation). Pennycook (2007) used the term translocalisation to account for how, with reference to hip-hop music, language moves across space and borders and still retains its local linguistic (and cultural) flavour; in other words, translocalisation consists of the appropriation and reinvention of English through several innovative strategies. Translocalisation is preferred to globalisation here because the discussion forum, Nairaland, has a particular audience and is visited mainly by Nigerians. Besides, there is a dominant use of linguistic features, borrowing from Nigerian languages and slang terms peculiar to Nigeria on the forum. Furthermore, the discussions on this forum involve local communities and a few participants in the diaspora through whom taboos are more widely disseminated.

Research Methodology

The data for this study were gathered mainly through interviews with 24 elders who volunteered to serve as informants for the research. The informants were from six different Yoruba communities stated below. The purpose of the research was explained to them and they were guaranteed anonymity. During the interviews, they were asked questions about Yoruba taboos, the historical backgrounds leading to some of the taboos, the roles of taboos in Yoruba land, the various types of taboos, their sociocultural values and how taboos could help to maintain peace, healthy living and cohesion among the Yoruba. The elders interviewed were over the age of 60. This was done to ensure that, given their experience and age, the information they gave about taboos was to a large extent authentic. Authentic information, in the context of this chapter, means facts and information that give the correct pictures of what constitute Yoruba taboos, their various types and the various ways in which they are used. It is expected that given their age, these elders are experienced enough to have adequate knowledge of Yoruba taboos.

For the purpose of this research, we visited the following towns: Owo in Ondo state, Igede-Ekiti in Ekiti state and Ara, Ede and Oba-Ile in Osun state. Ile-Ife, the cradle of Yoruba land and also where I live, is one of the towns where the interview was conducted. The distribution of the informants was uneven as participation depended solely on the willingness of these elders. Therefore, we interviewed three people in Owo, four in Igede-Ekiti and three in Ara. In Ede, four people participated in the research, three in Oba-Ile and seven in Ile-Ife, my city of residence. The duration of the interview with each of the participants was about 30 minutes and their responses were tape-recorded for analysis. We also made sure that all the participants were indigenes of the town where they were interviewed.

In addition to this primary source, data were also drawn from posts about taboos on Nairaland. We also sourced information about some of the taboos used in this study from the internet. In order to obtain additional data on taboos from Nairaland, we googled 'Yoruba taboos on Nairaland'. This search attracted more than 70 hits, although most of them were not relevant as they were not about Yoruba taboos. We then concentrated on three hits that were indicated not only to be from Nairaland but also on Yoruba taboos. The hits are 'What are cultural taboos in your village or area?', '10 cultural taboos in Yorubaland' and 'the taboo in your community'. It is from these that some Yoruba taboos were selected to complement the ones gathered from the interviews with our subjects.

Data Analysis and Discussions

The data gathered demonstrated that Yoruba taboos can be categorised into different types. Prominent in our data are (a) myth-related taboos, (b) lineage-related taboos, (c) religion-related taboos and (d) instructive taboos. However, these categories are fluid and leaky as sometimes there is no neat line of demarcation among them. Each of these categories is discussed below.

Myth-related taboos

Myth-related taboos are usually taboos that emanate from certain historical contexts relating to particular towns, villages, cities or communities (Akande, 2013). A myth-related taboo may stipulate that the king of a community (and in most cases whoever hails from that community) is forbidden to do or eat certain things. In Yoruba land, most traditional towns and rural areas have their own taboos that they must not violate. Some examples of such taboos are presented here as examples.

The first myth-related taboo is related to Ọ̀wọ̀, an ancient town in the Ondo state of Nigeria. The people of Ọ̀wọ̀, including the king, must never eat 'ẹdun' (monkey) meat. This is often couched in the saying 'A bími lỌ̀wọ̀, mi ò gbọdọ̀, jẹran ẹdun' ('I hailed from Ọ̀wọ̀ and so I must not eat monkey'). There is a historical background to this taboo. When Ojugbelu Arere, the first king of Ọ̀wọ̀, left Ilé-Ifẹ̀ with some people to settle elsewhere, they got lost in a thick forest and were stranded for days. According to the story, a monkey they met in the forest helped them by showing them the way to the place where they are settled today. So, to honour the monkey for the assistance it rendered, for they could all have died in that thick forest, they made a covenant with the monkey that no indigene of Ọ̀wọ̀ would ever kill or eat monkey (Oluwole, 1992; Owoseni & Olatoye, 2014). And today, there is always a monkey in front of Olowo palace to remind the indigenes of their forefathers' covenant with the monkey.

Eating dog meat is forbidden for any indigene of Ara, a town in Egbedore local government area of Osun state in Nigeria. This taboo is related to the abovementioned one. The founder of Ara was a great hunter whose means of livelihood was to go hunting for three or four months or even longer, kill game and sell it to people. On one of his hunting escapades, he got sick and died in the forest. It was his dog that came home, alerted the community members and actually took them to the scene of his death. If not for his dog, nobody would have known and the carcass would have rotted in the bush. So, the community members brought the remains of their leader home and gave him a befitting burial. Since that day, they view dogs as their friends whose meat must not be eaten. Several myths that forbid certain people from eating certain meats abound in Yoruba land. These two examples are similar to what Ajibade (2005) and Owoseni and Olatoye (2014) have pointed out in relation to the devotees of Oya: they must never eat buffalo meat.

In Ede, another ancient town in Osun state, it is taboo for the king, the Timi of Ede, to carry a young baby. There was a time in the history of the Yoruba kingdom when the Fulanis raided and looted many Yoruba towns. It was very difficult to raid Ede because the then Timi was a warrior. The Filanis came to him, gave him one of their beautiful daughters in order to know his secret. The king fell in love with the girl and revealed his secret – the source of his power was a pot of a herbal concoction that must not be broken. His wife also observed that the king loved children very much.

His wife revealed the secret to her people and so they planned to raid Ede. On the fateful day when the plan was to be carried out, his wife pretended that she was sick and asked the king to carry their young baby so that she could rest. While the king was playing with the baby, his wife went inside the secret room and broke the concoction pot. She immediately alerted her people and they raided Ede with the aim of capturing the king and looting the town. The king, with the baby in his arms, went straight to where the pot was and discovered that his 'beloved' wife had betrayed him by breaking the pot. Wanting to avoid being captured alive, Timi ran away with the baby strapped to his back. When he got to Agbale, a very popular area in the outskirts of the town, he imploded with the baby into the earth's crust. Several studies have been conducted on myth-related taboos and superstitions of specific towns (Akande, 2013; Odejobi, 2013; Oluwole, 1992). The people then reasoned that if it were not for his love for babies, the king could still have faced the Fulanis and fought them. From then on, it became taboo for any Timi of Ede to carry a baby.

In the Igede-Ekiti headquarter of Irepodun/Ifelodun local government area of Ekiti state, it is taboo to shout around the Elemi River. In an interview with the Onigede of Igede-Ekiti on August 13, 2014, the following was reported in *The Sun* (2014):

> The monarch said that it is a taboo to *shout aloud unnecessarily around the Elemi River.* He added that in the olden days the river did give out chickens (both male and female) as gifts to people. He said there are places in the Onigede's palace where *no one is allowed to shout aloud.* He said anyone who violated such taboos were (sic) forced to pay for it by buying a he-goat which he or she would bring for sacrifice. Nowadays, such taboos are seen as superstitions and not hearkened. (Emphasis mine; see http://sunnewsonline.com/new/?p=76782)

Lineage-related taboos

Among the Yoruba, there are different lineages and each lineage has its own taboo(s). There are certain items that members of a particular lineage are not supposed to eat, there are things that they are not supposed to say and certain behavioural patterns that they are not supposed to engage in (see Babalola, 1967). In traditional settings, members of a lineage are often informally taught the taboos that relate to their families so that they will not violate them. Any attempt to violate such taboos can lead to sickness, poverty, ostracism and even death. Some examples of lineage taboos in our data follow.

Onikoyi is a powerful lineage in Yoruba land; their 'oríkì', panegyric poetry, shows that they are great warriors whose business it is to protect the Yoruba kingdom. An extract from the panegyric poetry of Onikoyi in which their taboos are stated is

Ọmọdé Oníkòyí, Àgbààgbà Oníkòyí	The children of Oníkòyí , the elders of Oníkòyí
Wọn kì í jòkété	They must not eat bush rat
Nítorípé Atúgun ni baba wọn fi í ṣe.	As their forefathers used it as a charm to dispel wars
Bẹ̀ẹ̀ni Ìyàwó Oníkòyí	**And Oníkòyí 's wives**
Wọn kì i pAgbọ̀n lÁgbọ̀n	**They must not call a basket by its name**
Níjọ́ wọ́n bá pAgbọ̀n lÁgbọ̀n	The day they dare call a basket by its name
Wọn a rán Ọkọ wọn	They send their husbands
Síbi Ogún gbé le koko	To the tough warfront

Whoever is from the lineage of Onikoyi is forbidden to eat 'òkété' (bush rat) because it is one of the ingredients that their forefathers used as a war charm. Onipede and Adegbite (2014) as well as Owoseni and Olatoye (2014) also provide an account of lineage-related taboos and confirm that the lineage of Onikoyi is forbidden to eat bush rats. It is also forbidden for any wife of the lineage of Onikoyi to call a basket by its name (i.e. agbọ̀n). This means that Oníkòyí's wives must coin a term to refer to 'agbọ̀n' (i.e. basket) in Yoruba so that they do not send their husbands to the war front.

Similarly, the Oluoje lineage must never partake in the eating of a bird called 'Ègà'. When found, Ègà birds are usually in large colonies. It was popularly believed that when this lineage's enemies attempted to capture them by surprise, it was the Ègà, domiciled in one of the trees in Oluoje town, that saw the enemy coming and alerted the lineage. So, before their enemies arrived, the Oluoje were fully prepared and faced and defeated them. As a result, a covenant was made that Oluoje's descendants would never eat this bird. So, whenever anybody from the Oluoje's family eats the bird either intentionally or by accident, the consequences can be disastrous unless the ancestors are pacified.

In the same vein, the Obedu lineage, the majority of whose members are domiciled mainly in Oba-Ile in the Olorunda local government area of Osun state, is forbidden to eat 'Eku Àgó' (striped rat). An extract from the lineage 'oríkì' is

Òbà ò jàgó	Òbà does not eat Àgó rat
Ẹ ma fàgó rẹmí lẹkún	Do not soothe me with Àgó rat
A bi n lÓbà	I hailed from Òbà
Ta ní jẹ fẹran fínfín kojú Ọba?	Who dares give a striped rat's meat to the King

An interview with the Balogun of Oba revealed that if any member of the Obedu family eats Ago rat, such a person may die mysteriously (Akande, 2013). If this taboo is accidentally violated, the person who breaks the taboo will have to buy many items as a sacrifice.

Religion-related taboos

In Yoruba land, there is a multiplicity of traditional religions as each of the 401 deities is associated with certain religious practices as noted in Alade (2007). What this means is that the worshippers of each deity have certain taboos that they are supposed to observe and if they fail to do so, they may incur the wrath of their god. For instance, an Obatala worshipper in Yoruba land is not allowed to drink palm wine (Lawal, 1997). This taboo is associated with the belief that the primordial Obatala, whose business it was to mould human beings, drank palm wine one day when he was moulding human beings and made mistakes; this resulted in the creation of the physically or visually challenged human beings that we have today. So, in order to avoid making grievous mistakes, Obatala worshippers are generally prohibited from drinking palm wine. Similarly, Obatala worshippers are banned from eating red meat and they cannot wear any cloth that has a red colour (see Alade, 2007; Lawal, 1997). They must always be dressed in white as it is believed that white was the favourite colour of Obatala when he was alive.

A popular saying in Yoruba is 'Awo ki I san bante awo' ('an initiate does not wear the skirt of another initiate') which means 'an initiate must never betray another initiate'. A very important taboo in Yoruba land is that a traditional priest cannot marry or covet the wife of another traditional priest. This is very common among hunters. It is forbidden for a hunter to have any illicit relationship with the wife of another hunter. Whenever this happens, Ogun, the god of iron, can strike the culprit dead. Similarly, it is taboo to pour palm-kernel oil on Esu, which is often represented by yangi (sedimentary rock). If palm-kernel oil is poured on Esu and his spirit is invoked by mentioning the supposed (and sometimes fictitious) pourer, something very sinister will happen to the assumed pourer. It is also taboo for worshippers of the goddess Oya, the deity of the Niger River, to eat sheep meat because Oya hated sheep when she was alive. Familusi (2012) highlights various religious taboos that prohibit women from doing certain things during certain religious festivals. It is also believed that the worshippers of Oya must not eat pumpkin (Owoseni & Olatoye, 2014). A taboo related to Osun, a goddess of rivers, is that her worshippers must not eat guinea corn. Yoruba taboos are sometimes recorded on Ifa websites and Facebook pages. For instance, the Ifa verse known as Ogbedi is found on the Facebook page 'OyekuOfunTemple' (OyekuOfunTemple, n.d.) as follows:

Òní wiriwiri Ogbèdi	Today Ogbedi appears
Ọ̀la wiriwiri Ogbèdi	Tomorrow Ogbedi appears
Ẹyẹ nlá ní fapá sọ̀ gán	A big bird carries her arms at once
Òkú ọ̀pẹ̀ ní homù lẹ́yìn	A dead palm tree is one on whose back omu grows
A dífá fún Jẹnjòkè (Osun's hunter)	Cast Ifa for Jenjoke
Tí í ṣe wọléwọ̀de Òsun	The closest servant to Osun
Òsun ò màmà jẹ bàbà	*Osun did not eat guinea corn*
Ẹ̀yin ò ri bí bàbaà ṣe n wọnu Òsun	*Don't you see how it got inside her*
N kò màmà jẹun èèwọ̀ mọ́	*I shall never eat forbidden things*
Ó tún dayé àtúnwá mí	*Never in my life time*

Osun knew that she was forbidden to eat guinea corn. However, she did not know that the animals her hunter was killing and she was eating were killed in guinea corn farms. So, she was indirectly eating guinea corn and she eventually became barren. It was only when she stopped eating these animals after Ifa revealed the consequences of doing so that she became pregnant and had children. That this Ifa verse and its English translation are found on Facebook demonstrates that even non-Yoruba people can have access to it and understand the taboo it contains. Thus, in this respect, Facebook has contributed to the promotion of the Yoruba culture.

Finally in this section, the taboo that forbids Alaafin of Oyo to see the Elegun Sango of Koso face to face; it must never happen. Sango was once a king in the old Oyo empire and he is generally accepted to be the father of any Alaafin. However, the Elegun of Koso, Kòso being the place where Sango was said to have hanged himself, also represents the physical and supernatural power of Sango. Both Alaafin and the Elegun Sango are perceived to be one and it is not possible for either Alaafin or the Elegun Sango to bow to each other. So, whenever the Elegun Sango goes to the palace of Alaafin (and this usually happens once a year during the performance of traditional rites for Alaafin's crown), the Alaafin must go inside as he must not set eyes on him.

Instructive taboos

As hinted earlier, taboos constitute an important aspect of indigenous education (see Agboola & Mabawonku, 1996). Through mastery of taboos, a child (an adult and even a king) may have a good knowledge of what he/she is not supposed to do, what is expected of him/her in society and how he/she is supposed to relate with others.

There are several examples of indigenous education taboos but within this broad category, there are other subcategories. For instance, there are taboos that have to do with royalty. As an example, among the Yoruba, a king must not set eyes on a corpse. Though a king may attend a funeral ceremony, he can only be present after the corpse has been buried. It is also a taboo in Yoruba for all kings, except 'Ologotun omo Asijuwade' (king of Ogotun, he who can see the inner part of his crown), to see the inside of the crown on his head. It is an abomination which, if done by any king, such a king will be blinded for life.

Women are very powerful in Yoruba land and they play very crucial roles in checking the excesses of a despotic king. Whenever there are problems leading to the death of many people in a Yoruba community and it is obvious that the king is not doing enough to put a stop to the problems, the women in the community can be mobilised by their leader to visit the palace. On their way to the palace or even when they get there, they will turn their dresses inside out. It is forbidden for any Yoruba king to see women dressed this way. If any king sees women with their dresses turned inside out, there will be grievous implications for such a king. So, Yoruba kings are always mindful of this and they dare not incur the wrath of the women in their communities. Related to this is the taboo that 'Oba ko gbodo si igba' ('a king must never open the calabash'). Whenever a Yoruba king is rejected by his community for engaging in one atrocity or another, he can be compelled to open the calabash. There is always a traditional calabash in the custody of any Yoruba king; this calabash contains items that no king must see and it is usually covered. When a king becomes oppressive and is no longer loved by his community, the traditional chiefs will advise him to behave like a man by opening the calabash. The moment he opens it, he will join his ancestors. These are some of the ways through which the Yoruba exercise some control over their kings (Onadeko, 2008). Onadeko (2008) remarks that asking a king to open a calabash is a euphemistic way of asking him to commit suicide. So, princes in a typical Yoruba palace are often informally taught these taboos and many other related taboos so that they will have mastered them by the time they grow up.

There are also taboos that are meant to teach children how to respect their elders and, in most cases, taboos of this nature are found in proverbs. Aboluwadi (2014) has examined how African proverbs and the taboos in them can be used as teaching material in the African setting. An example of such a taboo is that when an elder and a younger person are eating from the same plate, the younger person is not allowed to be the first person to take a piece of meat. Also, when two elders are conversing, a child is not supposed to interrupt; this taboo is more or less a mark of respect for elders.

There are Yoruba taboos that express their world views. A father is prohibited from inheriting his son's property, including his wife. It is believed that, naturally, a father should die before his child and it is the child who should inherit from his father and not the other way round. So, when the contrary happens, it becomes anomalous for the father to take possession of his child's property (Odejobi, 2013). Another taboo in Yoruba land is that when an 'okete' (a bush rat) stands on two legs, it must not be killed. It is believed that if it is killed while standing on two legs, the killer will lose any future child born to him. It is also forbidden for a child to fall from her/his mother's back. If this happens, it is believed that, if the child is female, she will first lose seven husbands and only her eighth husband will live long; if the baby is male, he will lose nine wives until he marries the tenth one. Although the examples are inexhaustible, the point to foreground here is that the taboos that fall under this category have some connection with culture, experience and tradition.

Yoruba Taboos on Nairaland

In this section, we highlight some of the important Yoruba taboos on the website Nairaland. These are general taboos that apply across all Yoruba land and are familiar to every Yoruba person. It is important to state here that the posts are deliberately not edited; they appear exactly the way they are posted on the forum. The first example (Example 1) is one of the posts in the thread 'What Are Cultural Taboos In Your Area?' while the other examples (Examples 2 through 5) are posts in another thread titled '10 Cultural Taboos in Yorubaland.'

Example 1
There's a masquerade in Ibadan, Oloolu, it is said that women are not allowed to see this masquerade...any woman that does this will be drained of her blood instantly Rumor has it that a childless woman deliberately crossed his path one year seeking to die cos she is childless but irrespective of that the thing took pity on her and gave her a baby. I dunno the authencity of the rumor though But I know for sure women are not allowed to see it, ever. You see women running Helter Skelter trying to escape its path ('What Are Cultural Taboos in Your Village or Area', n.d.)

Example 2

The surest death penalty is when a woman set her eyes on oro ('Nairaland Forum', n.d.)

Example 3

If someone dies in a river, he must be buried beside the river. His corpse must never be returned home for burial to avoid mass death in that house. Even the whites know about this, yeah, Mungo Park.

Example 4

The DunDun Drum also known as Gangan a very popular two-sided drum in yoruba land. Very popular for its unique sound, it can be used to pass across messages (talking drum). But, no matter how angry or offended the hearer is about the message being passed across, he must never pierce this drum with a knife in the midst of a crowd or gatherings. The one who does will end up with diseases and sicknesses that will eventually end his life.

Example 5

Seeing a mouse (okete/ewu) in the afternoon, Yoruba believes something tragedy is going to happen if you see a mouse in the afternoon.thats why they say 'a n kin ri ewu losan' ('10 Cultural Taboos in Yorubaland', n.d.)

In terms of language use, it can be observed that some of the features that have been identified by scholars as characterising CMC are present in the above examples (see Androutsopoulos, 2011a; Lemke, 1998). In Example 1 for instance, punctuation marks are not used at all in certain contexts where we would expect them. After the words 'instantly' in Line 2 and 'though' in Line 5 of the example, one would expect to see a full stop. Also noticeable is the wrong use of capitalisation as evident in the spelling of the words 'helter skelter' where the letters 'h' and 's' are improperly capitalised. Similar to this, is the use of what has been termed electronic spelling (see Akande & Akinwale [2010] on spelling practices in text messages). Evidence of electronic or phonetic spelling is illustrated in the spelling of 'because' and 'do not know' as 'cos' and 'dunno', respectively. Obviously, the spelling of words in the digital space does not normally follow the conventional standard spelling norm; words are orthographically shorter in digital space than they are in traditional writing. The electronic spelling identified here is one of the features of Netspeak identified by Crystal (2004: 81–93). He remarks that for economy, expressions like 'be back later', 'as soon as possible', 'call for comments' and 'great' can be written as 'bbl', 'asap', 'cfc' and 'gr8', respectively. Crystal (2004: 88) observes that 'Non-standard spelling, heavily penalized in traditional writing (at least, since the eighteenth century), is used without sanction in conversational settings' in digital spaces. There are also instances of code-mixing as Yoruba words such as 'oloolu' in Example 1, 'oro' in Example 2, 'gangan' in Example 4 and 'okete' in Example 5 were used by discussants. The use of code-mixing and borrowing confirms earlier studies carried out by Androutsopoulos (2011b) and Siebenhaar (2006) on CMC.

A taboo that is very common in Ibadan, a cosmopolitan city in Oyo state of Nigeria, is found in a post on Nairaland by MsNas. In this example (i.e. Example 1), the post states that it is a taboo that a woman must not come in contact with Oloolu, one of the most popular masquerades in Ibadan. If by any stroke of misfortune, a woman meets Oloolu, the woman must die instantly. It is also forbidden for a woman to see Oro. In Yoruba land, Oro is a cult that no woman, however highly placed or influential, can participate in or be a member of (Simola, 1999). Akanji and Dada's (2012: 20) study corroborates this claim. They also remark that the Oro cult has dignitaries, such as kings, chiefs and eminent men in society as its members. Whenever Oro is out (in most cases the outings happen at night, except in places like Ikorodu and Ijebu where they are during the day), because it is a male-dominant cult, no woman is allowed to see it. If a woman does see Oro, she must die immediately. This is why the Yoruba often say 'Awo Egungun lobinrin le se, awo Gelede lobìnrin le mo, bi obinrin ba foju kan Oro, Oro yo gbe' ('a woman can belong to the cult of Egungun, a woman can have the knowledge of Gelede but if a woman set her eyes on Oro, her doom must surely come').

Except for the taboo in Example 1, which is city specific, all the other taboos in this section are general. As we can see in Example 3, it is forbidden in Yoruba land to bring home the corpse of anybody who dies in a river; the remains of such a person must be buried beside the river after a series of sacrifices to appease the goddess of the river have been offered. The taboo in Example 4 is also general in that the drum, gangan, that is being referred to is popular in most Yoruba towns and settlements. An attempt by an individual to destroy this drum by piercing it with a knife can incur the wrath of Ayangalu the consequence of which may be instant death or severe sickness. It is also a taboo in Yoruba land for anybody to see okete (a bush rat) in the afternoon. Seeing a bush rat in the afternoon is a bad omen and when this happens, something tragic will happen to the person who beholds it. Examples 1 through 5 were all extracted from Nairaland, which, as hinted at earlier, is accessible to both Nigerians and foreigners. Through this digital forum, foreigners can obtain a good knowledge of many Yoruba taboos, and they sometimes ask questions relating to the rationale of certain taboos, the implications of violating the taboos as well as the socio-religious context that surrounds them.

Conclusion

This chapter set out to examine the social cultural values of Yoruba taboos among the Yoruba of the southwestern part of Nigeria and to provide a broader classification of these taboos. The chapter demonstrated that taboos are often used to control people's behaviour and guide the way that they relate with other people in society. As shown in this chapter, in traditional Yoruba society, the observance of taboos is one of the chief means of maintaining harmony and a peaceful coexistence. It was also revealed that taboos are contained in other cultural resources such as Ifa verses, oriki, proverbs and incantations.

The taboos discussed in this chapter were classified into broader categories based on historical, religious and functional contexts. The broad categories that this chapter has therefore come up with are:

(a) myth-related taboos;
(b) lineage-related taboos;
(c) religion-related taboos;
(d) instructive taboos.

Most importantly, the chapter emphasised the fact that Yoruba taboos are no longer purely oral in mode because they have been digitised. The chapter showed that Nairaland (an online discussion forum), Facebook and online news now serve as information sites where taboos are recorded and disseminated. The discussions of taboos in the digital age were presented to have some implications on language use as evident by the presence of such CMC linguistic features as phonetic spelling, code-mixing and non-use or haphazard use of punctuation marks.

The chapter concluded that many of these taboos are being translocalised, since the majority of them now transcend the traditional oral space, and increasingly occupy the digital space. The current advancement in digital and technological mediums has now made it possible for Yoruba traditional orature, of which Yoruba taboos constitute a part of, to be accessed by people from different parts of the globe.

References

'10 Cultural Taboos in Yorubaland' (n.d.) Nairaland. See www.nairaland.com/1927574/10-cultural-taboos-yorubaland-things#26759710 (accessed February 2017).

Aboluwodi, A. (2014) Exploring African proverbs as a learning resource in the contemporary society. *Research on Humanities and Social Sciences* 4 (17), 33–39.

Adebileje, A. (2012) Socio-cultural and attitudinal study of selected Yoruba taboos in southwest Nigeria. *Studies in Literature and Language* 4 (1), 94–100.

Afe, A.E. (2013) Taboos and the maintenance of social order in the old Ondo province, southwestern Nigeria. *African Research Review* 7 (1), 95–109.

Agboola, T. and Mabawonku, A.O. (1996) Indigenous knowledge, environmental education and sanitation: Application to an African city. In D.M. Warren, L. Egunjobi and B. Wahab (eds) *Indigenous Knowledge in Education* (pp. 78–94). Ibadan: University of Ibadan Press.

Ajibade, G.O. (2005) Animals in the traditional worldview of Yoruba. *Folklore* 30, 155–172.

Akanbi, T.A. (2015) Vulgarity in Yoruba proverbs: Its implications and sociological effects. *Studies in Social Sciences and Humanities* 2 (1) 173–181.

Akande, A.T. (2012a) *Globalization and English in Africa: Evidence from Nigerian Hip-Hop.* New York: Nova Science.

Akande, A.T. (2012b) The appropriation of African American vernacular English and Jamaican Patois by Nigerian hip-hop artists. *Zeitschrift fur Anglistik und Amerikanistik: A Quarterly of Language, Literature and Culture* 3 (3), 237–254.

Akande, A.T. (2014) Hybridity as authenticity in Nigerian hip-hop lyrics. In V. Lacoste, J. Leimgruber and T. Breyer (eds) *Authenticity: A View from Inside and Outside Sociolinguistics* (pp. 267–284). Berlin: Mouton de Gruyter.

Akande, A.T. and Akinwale, O.T. (2010) Spelling practices in text messaging. In R. Taiwo (ed.) *Handbook of Research and Discourse Behaviour and Digital Communication: Language Structures and Social Interaction* (pp. 349–363). Hershey, NY: IGI Global Publishing.

Akande, I.D. (2013) Agbeyewo Litireso Alohun Awon Orisa Ajemogun ni Awon Ilu Ajoruko-mo-Obani Ile Yoruba [An examination of war-related deities in Oba-named towns in Yoruba land]. Unpublished MA thesis, Obafemi Awolowo University, Ile-Ife.

Akanji, O.R. and Dada, O.M.O. (2012) Oro cult: The traditional way of political administration, judiciary system and religious cleansing among the pre-colonial Yoruba natives of Nigeria. *The Journal of International Social Research* 23 (5), 19–26.

Alade, A.L. (2007) *Taboos and Superstition.* Lagos: Alpha Press.

Androutsopoulos, J. (2011a) Language change and digital media: A review of conceptions and evidence. In T. Kristiansen and N. Coupland (eds) *Standard Languages and Language Standards in a Changing Europe* (pp. 145–159). Oslo: Novus Press.

Androutsopoulos, J. (2011b) From variation to heteroglossia in the study of computer-mediated discourse. In C. Thurlow and K. Mroczek (eds) *Digital Discourse: Language in the New Media* (227–298). New York: Oxford University Press.

Appadurai, A. (1990) Disjuncture and difference in the global cultural economy. *Public Culture* 2, 1–24.

Babalola, A. (1967) *Awon Oriki Orile.* Glasgow: Collins.

Bell, C. (1987) Ritualization of texts and the textualization of ritual in the codification of Taoist liturgy. *History of Religion* 27 (4), 366–392.

Carpenter, D. (1992) Language, religion, and society: Reflections on the authority of the veda in India. *Journal of the American Academy of Religion* 60 (1), 57–77.

Crystal, D. (1965) *Linguistics, Language and Religion.* Stroud: Hawthorn Books.

Crystal, D. (2004) *Language and the Internet.* Cambridge: Cambridge University Press.

Durkheim, E. (1963) *Incest: The Nature and the Origin of the Taboo.* New York: Lyle Stuart.

Familusi, O.O. (2012) African culture and the status of women: The Yoruba example. *Journal of Pan African Studies* 5 (1), 299–313.

Heyd, T. and Mair, C. (2014) From vernacular to digital ethnolinguistic repertoire: The case of Nigerian pidgin. In V. Lacoste, J. Leimgruber and T. Breyer (eds) *Authenticity: A View from Inside and Outside Sociolinguistics* (pp. 244–268). Berlin: Mouton de Gruyter.

Holden, L. (2000) *Encyclopedia of Taboos.* Oxford: ABC CLIO Ltd.

Ibagere, E. (2014) The media and the need to harness traditional taboos and injunctions for social cohesion in Nigeria. *International Journal of Social Sciences, Arts and Humanities* 2 (2), 22–32.

Idowu, B.E. (1962) *Olodumare: God in Yoruba Belief.* London: Longman.

Jegede, A.S. (2002) The Yoruba cultural construction of health and illness. *Nordic Journal of African Studies* 11 (3), 322–335.

Keane, W. (1997) Religious language. *Annual Review of Anthropology* 26, 46–71.

Lawal, B. (1997) *The Gelede Spectacle: Art, Gender and Social Harmony in African Culture.* Washington, WA: Washington University Press.

Lemke, J. (1998) Multiplying meaning: Visual and verbal semiotics in scientific text. In J.R. Martin and R. Veel (eds) *Reading Science* (pp. 87–113). London: Routledge.

Liman, A.A. (2010) Orature and multimedia: Exploring the inter-disciplinary potentials of an emergent field. *Journal of the Nigerian English Studies Association* 13 (2), 125–136.

Madu, S.N. (2002) Health complaints of high school students in the northern province and taboo themes in their families. *South African Journal of Education* 22 (1), 65–69.

McLuhan, M. (1962) *The Gutenberg Galaxy: The Making of Typographic Man.* Toronto: Toronto University Press.

Mills, A.K. (2010) A review of the 'digital turn' in the new literacy studies. *Review of Educational Research* 80 (2), 246–271.

Nairaland Forum (n.d.) Nairaland Forum. See www.nairaland.com (accessed February 2017).

Odejobi, C.O. (2013) An overview of taboo and superstition among the Yoruba of southwest of Nigeria. *Mediterranean Journal of Social Science* 4 (2), 221–226.

Ogunyemi, O. (2007) The implications of taboos among African diasporas for the African press in the United Kingdom. *Journal of Black Studies* 10 (10), 1–22.

Olabode, B.O. and Siyanbola, S.O. (2013) Proverbs and taboos as panacea to environmental problems in Nigeria: A case of selected Yoruba proverbs. *Journal of Arts and Contemporary Society* 5 (2), 56–66.

Olarinoye, A.O., Adesina, K.T., Olarinoye, J.K., Adejumo, A.O. and Ezeoke, G.G. (2014) Food taboos among pregnant Nigerian women. *Centrepoint* 20 (1), 12–26.

Oluwole, S.B. (1992) *Witchcraft Reincarnation and the God-Head.* Lagos: Academy Press.

Onadeko, T. (2008) Yoruba traditional adjudicatory systems. *African Study Monographs* 29 (1), 15–28.

Onipede, K. and Adegbite, F.A. (2014) Igbon, Iresa and Ikoyi: A pre-historic relationship till present time. *Historical Research Letter* 15, 24–27.

Osei, J. (2006) The value of African taboos for biodiversity and sustainable development. *Journal of Sustainable Development in Africa* 8 (3), 42–61.

Owoseni, A.O. and Olatoye, I.O. (2014) Yoruba ethico-cultural perspectives and understanding of animal ethics. *Journal for Critical Animal Studies* 12 (3), 97–118.

'OyekuOfunTemple' (n.d.) Oyeku Ofun Temple. See www.facebook.com/OyekuOfunTemple/posts/576497989076808 (accessed February 2017).

Pennycook, A. (2007) *Global Englishes and Transcultural Flows.* London: Routledge.

Siebehaar, B. (2006) Code choice and code-switching in Swiss-German internet relay chat. *Journal of Sociolinguistics* 10 (4), 481–509.

Simola, R. (1999) The construction of a Nigerian nationalist and feminist: Funmilayo Ransome Kuti. *The Nordic Journal of African Studies* 8 (1), 94–114.

Terantino, J. and Graf, K. (2011) Using Facebook in the language classroom as part of the net generation curriculum. *The Language Educator*, 44–47.

The Sun (2014) Igede-Ekiti: A town steeped in myths, taboos. See Igbokwenuradio.com/igede-ekiti-a-town-steeped-in-myths-taboos (accessed 14 May 2016).

Thody, P. (1997) *Don't Do It. A Dictionary of the Forbidden.* London: The Athlone Press.

Thorpe, C.O. (1972) *Àwon èèwò ilè Yorùbá.* Ibadan: Onibon-Oje Press.

Tomlinson, J. (1999) *Globalization and Culture.* Cambridge: Polity Press.

Wallace, A.F.C. (1966) *Religion: An Anthropological View.* New York: Random House.

Wardhaugh, R. (2006) *An Introduction to Sociolinguistics.* Oxford: Blackwell.

'What Are Cultural Taboos in Your Village or Area' (n.d.) Nairaland Forum. See www.nairaland.com/1540325/what-cultural-taboos-village-area#19949053   (accessed February 2017).

Wheelock, W.T. (1982) The problem of ritual language: From information to situation. *Journal of the American Academy of Religion* 50 (1), 49–71.

6 Yiddish Wikipedia: History Revisited

Tatjana Soldat-Jaffe

Not only is Wikipedia an open forum, visible and accessible to everyone, but it also facilitates interactions. As such, it not only aids dialogue but it also demands various degrees of collaborative effort to establish relationships between users that imposes certain demands on the individuals involved; it creates a social space for this virtual community. Yiddish Wikipedia as a virtual community becomes an extension of the diaspora community, which is, by definition, a displaced community and contingent on the relationship the respective users assume with each other as well as with the competitive Hebrew Wikipedia. Studying the entries as well as the discourse on the history page shows how conflicting interests develop into overt social conflicts. Ultimately, it appears that intergroup competition within Hebrew Wikipedia users enhances intergroup morale, cohesiveness and cooperation. Study of the Yiddish Wikipedia intergroup behavior in the form of perceived ethnic identity, on the one hand, and interpersonal behavior in the form of agentive subjectivity in a virtual community, on the other, are responsible for two types of conflict: the objective (i.e. factual) disagreement that is actually a subjective disagreement, and the explicit conflict that is actually of a long-standing historical nature.

What is Wikipedia? A Quick Introduction into the Quirks of Wikipedia

Baker (2008) observed in the *New York Review of Books* that '[m]ore people use Wikipedia than Amazon or eBay – in fact it's up there in the top-ten Alexa rankings with [...] MySpace, Facebook, and YouTube', or as Rosenzweig (2006) put it two years earlier, '[t]he Alexa traffic rankings put it at number 18, well above the New York Times (50), the Library of Congress (1,175) and the venerable Encyclopedia Britannica (2,952)'. Although there no current statistics are available, we can expect that the number of users has continued to rise. Wikipedia is a web-based free encyclopedia project launched in January 2001 and operated by the Wikimedia Foundation, a nonprofit charitable organization.[1]

According to its founder Jimmy Wales (2002), the goal of Wikipedia is to 'give every single person in the world free access to the sum of all human knowledge'. Since the beginning of Wikipedia in 2001, the number of users has steadily increased; more often than not, Wikipedia has assumed the role of the sole source of reference as a free encyclopedia and has pushed more conventional print sources into the background. Ultimately, Ayers *et al.* (2008: 407) points out, '[w]ikipedias have been created in over 250 languages, each representing its own individual community and unique collection of content'.

Yiddish Wikipedia is no exception to this trend. It is well aware of its presumed authority as a definitive online resource: a peer collaboration on projects where written works do not have owners but have multiple anonymous authors. As an online resource, it takes advantage of its easy access, allowing it to be widely read and cited. However, adopting Yiddish Wikipedia as an authoritative site that compiles and engages historical 'knowledge' becomes, as this chapter will demonstrate, a challenge for Jewish studies scholars in general, and for Yiddish scholars and community members in particular, because a clear distinction between fact and opinion is often difficult to draw.

Because Yiddish emerged as a diaspora language from a history of religious, cultural and linguistic wars, and for what is today a hybrid community, many discussions on Yiddish Wikipedia exhibit how past experiences heavily determine the local dominant discourse. Yiddish Wikipedia as a virtual community becomes, as will be shown, an extension of the diaspora community which is, by definition, a displaced community and contingent not only on its relationship with the respective contact cultures but also in constant negotiation (or lack thereof) with the Hebrew-speaking community.

To push this argument further, the language practice, that is, the recurrent pattern of language behavior, on the Yiddish Wikipedia site reflects Yiddish's long-standing cultural struggle in its demand to find a co-existence with Modern Hebrew. The purpose of this chapter is to illustrate the relationship between language practices displayed on Wikipedia and the participants' membership affiliation (Yiddishist versus Hebraist) in which language practice becomes essential to the development of a distinctively political language psychology. As I argue, the respective language practices are politically charged so that they implicitly and explicitly aim to manufacture as well as 'correct' perceptions, and, by extension, confirm or alter the historical narrative by linking individuals' internal reality to historical cultural wars between Yiddish and Modern Hebrew.

The Jewish Dispute

Yiddish's history (cultural as well as linguistic) has been marked by life in diaspora. More often than not, Yiddish has had to define itself against its contact languages – whether languages in its immediate environment (such as English, German, French) or the other Jewish language (i.e. Hebrew). Its difficult and tense relationship with Modern Hebrew is well documented.[2] As a diaspora language *par* excellence – perhaps, the exemplary language of the diaspora – Yiddish has been in constant struggle with multiple shifting environments, whether these be its many non-Jewish contact cultures without or assorted Jewish cultural contexts within. These and other historical exigencies have given the idea of the language a particular set of cultural meanings, what Benjamin Harshav has aptly described as the 'semiotics of Yiddish'. In the 19th century, Yiddish went from being largely taken by proponents of Jewish enlightenment as an outmoded obstacle to emancipation and cultural assimilation to having its Jewishness denigrated by Zionists who married Jewish national destiny to a renascent, secular Hebrew. The diversity of its contacts, shifting and uncertain national embeddedness and conflicting status ambitions predispose Yiddish to a distinctive and complex functional dynamism for self-reflexivity.

This leaves Yiddish with several unsettled issues about Jewish identity in modernity, and an antithesis by anti-Yiddishists, who maintained Yiddish as an inauthentic hindrance to group progress. The conflicted discourse plays out in two ways. First, it informs the set of constitutive and definitional arguments about the relation between the Yiddish language and Jewish identity in terms of religion, ethnicity and nationalism. Second, it underwrites the arguments about the status of the language as an agent of mediation in the Jewish experience of modernity.

The animosity between Hebraists and Yiddishists is historical and is best reflected in a vehement protest expressed by Chaim Zhitlovski (1904), an important Yiddish intellectual, who criticized the roles that had been issued to the two Jewish languages. Behind his rhetoric about present demographic realities for Yiddish – the mother tongue, called *mame-loshn* – lies a criticism of Hebrew – the holy language, also referred to as the *loshn-koydesh*:

> How can we say that the holy language is the language of the Jewish people? By no means! If that were the case, one would have to say that the national language is a language that the people do not speak and do not understand. The holy language is the language of our intellectuals, better said: the language of a fraction of our intellectuals. And as such not even a language for conversational purposes but rather language for writing just like Latin used to be once the written language of the educated people in the Middle Ages. (Zhitlovski, 1904: 122)

Zhitlovski's case relies on an argument based on modernity and actuality. Yiddish is not merely the majority language of ordinary Jews; it is also the present-day language of secular Jewish intellectual life. This move is a pivotal leap forward in the Yiddishism debate, shifting the emphasis away from the doctrine of linguistic purity that was the mainstay of anti-Yiddishists. For Yiddishism, the past and present dynamic – obsolescence versus modernity – is more important than the purity versus impurity dynamic (Zhitlovski, 1904: 122).

Zhitlovski's flanking maneuver was a performative declaration that Yiddish was more than a vernacular – a language of the Yiddish *gas* and *shtetl*. Yiddish, he claims, was the language of a vital intelligentsia. Hebraists and linguistic assimilationists alike made the case that Yiddish was a debased, even bastardized language, favoring the pejorative moniker *jargon* whenever possible to underscore this attribution. Hebraists were particularly partial to this line of thinking; Hebrew, the sacred language, is pure, masculine and uncorrupted; while Yiddish is fallen, feminine and somehow corrupted. The authority (and 'quality') of Biblical Hebrew (*loshn-koydesh*) has been extended to Modern Hebrew so that the same tension can be detected on Yiddish Wikipedia as will be seen in the examples from the Yiddish Wikipedia site. We see the same narrative and behavior in the defense of Yiddish Wikipedia: being accused of not being a proper language, not having enough speakers and not having authentic speakers.

The cultural war between Yiddish and Hebrew goes back to questions concerning Jewish identity, self-definition and Yiddish's demand for autonomy. I have argued elsewhere that the core of the dispute rests on the dual meaning of the word Yiddish (*yidish*). Simply put, in Yiddish, the word 'Yiddish' means both Yiddish and Jewish. In effect, this usage signals *yidishkeyt*, the specific form of Jewishness particular to Yiddish speakers (Soldat-Jaffe, 2012). Ultimately, for Hebraists, the definitional question – defining a single and singularly Jewish language – was foremost a national language question. Seemingly, a monolingual Judaism was a necessary condition for a presumptive Jewish nation state. The difficulty of defining Jewishness – slippage among notions of ethnicity, folk, nation and religion and the fact that the word *yidish* itself means Jewish – was a resource for Yiddishists. Questions of national language often revolve around accounts of historical pedigree and national history. In the case of Hebrew, these answers came from religious history, a seemingly unimpeachable source of authority for defining a language of Jewish nationhood, given the chief role of Hebrew (and Aramaic) in the liturgical and source texts of Judaism. Hebraists claimed authority by asserting their continuity with a source in the remote, albeit sacred, linguistic past. With a language ideology that prized religious authenticity and purity, Hebraists accused Yiddish of not meeting their definition of a proper Jewish language. In effect, claims for Hebrew as the Jewish language came from a creative appropriation and simplification of the notion of Jewish scriptures as a source of law – in this case denoting the source of a *de jure* argument about language. Yiddish Wikipedia contributions lay bare and act out these very debates, as they fight over Yiddish self-definition, rejecting the symbiosis and interdependence between Yiddish and Hebrew as two Jewish languages. As such, Yiddish Wikipedia – whether in support of or in opposition to Yiddish Wikipedia – represents a measure to defend, rescue and reinforce Yiddish identity and community while its inherent contentions seem to be putting Wikipedia's guidelines to a test.

Web Community as a Community of Practice

While MySpace, Facebook and YouTube represent the expression of individual viewpoints and opinions, Wikipedia, being a free encyclopedia, strives to be a reference source and, as such, seeks to be *factual*. This objective is written down in the general Wikipedia guidelines, which state that '[a]ll encyclopedic content on Wikipedia must be written from a *neutral point of view* (NPOV), which means representing fairly, proportionately, and, as far as possible, without bias [..]. NPOV is a fundamental principle of Wikipedia and of other Wikimedia projects'.[3] As Ayers *et al.* (2008: 37) conclude, this means that 'articles should be concise surveys, not personal essays [...] accurate and objective'.

And yet, Wikipedia's design has made it a challenge to pursue this objective. As a web community, Wikipedia is best defined as a social network with a common interest pursuing a common goal. While web community is used, here, interchangeably with the terms 'online community' and 'virtual community', it is necessary to understand what impact this information distribution has on its audience. According to Wenger *et al.* (2002), virtual communities are communities of practice in which members can share their mutual understanding of knowledge. These communities are characterized by their shared interest in a practice. As Ahuja and Carley (1999) point out, in a virtual community, the communication and coordination of work take place in cyberspace through information technology. Therefore, the coherence of a community is relational without reference to a specific location (Ahuia & Carley, 1999). As such, they defy traditional geographical boundaries and (must) impose a different organizational structure to make up for the absence of physical referentiality as they only make use of electronic forms of communication for exchanges. Sköld (2015: 296) points out in his research that the members of a virtual community, such as Wikipedia, are solely bound together by a commonality of interest. This approach resonates with Kumar *et al.*'s (1999) approach, which states that a web community is merely related to a topic as well as to the knowledge about this topic. *Ad rem*, the virtual nature of the web often makes the web vulnerable as the identity of the participant seemingly takes a back seat to the topic. It forms the illusion that the identity of the community is realized through its narrative.

Kumar *et al.* (1999) explicate in more detail the nature of the web, maintaining that there are two types of communities on the web: explicitly and implicitly defined communities. Explicit communities are communities that are easily identified on the web by an obvious membership such as the Automotive Club. In accordance with this theory then, implicit communities, such as Wikipedia, are comparatively more complex because they often refer to distributed, *ad hoc* and random content creation related to the common interests on the internet. While the explicit web community follows a top-down structure, the implicit web community is a bottom-up one. As such, to create a document, the implicit web community structures, navigates and negotiates not only information space but also social space. This, as it turns out, is both a virtue and a weakness. Although it dismisses any limitations on the size of a network and it allows unrestrained and unconditional access to the network, the management of it can become challenging; primarily, as it turns out, it becomes difficult to maintain two of Grice's (1989) maxims of conversations: the *maxim of quality* (be truthful and do not give information that is false or that is not supported by evidence) and the *maxim of manner* (be as clear, as brief and as orderly as one can in what one says, and where one avoids obscurity and ambiguity). Not only are agency (the capacity of acting or of exerting power) and accountability (taking responsibility for one's action) out of sync with participants, but agency often takes precedence over accountability as participants' profiles are not accessible and their integrity is often questionable.

Being a hybrid platform, Wikipedia is, on the one hand, an open forum, visible and accessible to everybody, but, on the other, it also facilitates interactions and serves foremost as a social space. As such, not only does it aid a dialogue between users but it also demands various degrees of collaborative effort to establish relationships between these users, which imposes a certain number of demands on the individuals involved. In other words, it expects the users to go beyond the plain discursive level and to take into consideration the surrounding social and historical context to create cohesion.

Busher and Nalita (2015) revisited the concept of a hybrid community, expanding on the definition initially provided by Bhabha (1994) by investigating in depth how people construct meanings through the relationships and communications they develop with others. Bhabha (1994: 4) claims in his notion of hybridity that there is a space 'in-between the designations of identity' and that 'this interstitial passage between fixed identifications opens up the possibility of a cultural hybridity that entertains difference without an assumed or imposed hierarchy'. While Bhabha's approach addresses the subject of colonialism and his understanding of hybridity relies heavily on ethnic identity, Busher and

Nalita (2015) understand identity and hybridity in a broader sense that is free of any particular markers. They argue that 'these are interconnected with the development of new or emergent communities, cultures, and personal identities constructed by their members' (Busher & Nalita, 2015). The authors insist that understanding the mechanics behind hybrid communities is important for understanding the participants' agency in an unknown and indefinable (virtual) environment:

> This [...] involves meeting with unfamiliar people in unfamiliar institutional spaces with whom they have to negotiate in unfamiliar socio-political contexts. [...] [U]nderstanding who are the 'others', the notion of alterity and who are the 'same' is important for helping people recognize with whom they have to interact dynamically to assert their agency. (Busher & Nalita, 2015: 169)

In other words, hybrid communities are not cohesive and homogeneous social objects as such; instead, membership in hybrid communities requires recurrent referentiality as these communities are dynamic and defy the notion of pre-established permanent parameters (i.e. ignoring parameters that are context dependent and deliberate as can be found in explicit communities).

Just as implicit web communities thrive on opaque membership identities, these communities also render a different type of outcome. While traditionally, a piece of information had to pass through several layers of weeding, selecting and editing before it would get a fighting chance to come to the attention of a wider audience, online publications can be contributed to without those built-in mechanisms. This is reflected, for example, in the use of hypertext in Wikipedia. Hypertexts give the encyclopedia a different organizational style as the linking extends beyond articles. It lets the reader select a specific discursive path that could potentially deviate from (expand or zoom in on) the given frame of information.

Because contributions and editing can be performed by everybody in an online reference source that is open to the public, as Ayers *et al.* (2008: xxi) point out, Wikipedia editors are in charge of updating or changing the encyclopedia whenever they see a problem to make sure that input is truly definitional. Looking at this Wikipedia problem from a software engineering point of view, Frederick P. Brooks' (1975; author of the classic book *The Mythical Man-Month*) 'Brooks' Law' comes to mind. Brooks talks about the inherent complexities of coordination and states that as the number of involved programmers N rises, the work performed also scales as N, but the complexity and vulnerability to mistakes rises as N-squared, in accordance with the number of communication paths required to coordinate the contributors. To achieve quality, Brooks (1975) therefore recommends a minimum of contributors: 'Conceptual integrity in turn dictates that the design must proceed from one mind, or a very small number of agreeing resonant minds'. Ford points out that a cooperative editing is often overshadowed by personal exercise of opinions. He maintains that "the deployment and removal of cleanup tags can be seen as an act of power play between editors [which] affects readers' evaluation of the article's content" (Ford, 2015). While it is not in Wikipedia's nature to observe Brooks' Law, it does understand the complications an open forum creates.

A weak attempt to tackle this problem, following Ayers *et al.*, comes in the form of two rules of conduct that Wikipedians, members of the Wikipedia community, are expected to follow. These are the 'Assume Good Faith' and 'Please Don't Bite the Newcomers' rules (Ayers *et al.*, 2008: 332).[4] However, as the authors point out, 'Assume Good Faith' and 'Conflict of Interest' often put a strain on the Wiki work since an editor's intention is not always transparent or 'in good faith'. To make sure that conduct does not become aberrant, governance is reassured through subcommunities called WikiProjects that act as local jurisdictions. Ultimately, Wikipedia's hybrid platform facilitates a hybrid governance that will respect Wikipedia's constitution. Ostrom (2000) explains the preference of an *internal* proportionate mechanism of policing in the form of intrinsic social norms as an effective voluntary social apparatus over a 'top-down' rule-imposed external authority: the former supports a more cooperative behavior. Ostrom (2000: 147) notes that not only are social norms equally effective at 'generating cooperative behavior' but also that 'moreover, norms seem to have a certain staying power in encouraging growth of the desire for cooperative behavior over time, while cooperation enforced by externally imposed rules can disappear very quickly'.[5]

Cooperative behavior is further strengthened by the support of interpersonal communication channels. One of the advantages of the social media revolution is the possibility to record and study information about interpersonal communication much more easily as it allows for personal online relationships that can be tracked via *talk page* (Ayers *et al.*, 2008: 263). Closely associated with this, clicking the history page of the talk page will lead you to the history of the talk page rather than the history of the article it is associated with. As Ayers *et al.* (2008: 113) point out, '[t]alk pages are important, socially and practically. They help strengthen content, and they're also an integral part of Wikipedia's community. [...] [T]hey function as a space for conversation between all the readers and editors of an article'. According to the authors, the importance of signing the comments on a talk page emphasizes the intimacy of this specific platform (Ayers *et al.*, 2008: 116).

The availability of transparency is another way to boost cooperative behavior. Wiki editing can be seen from the start as a way to share knowledge and to participate in a continuing conversation among peers with the goal to establish a permanent encyclopedic entry. Participation in a continuing conversation can be followed on the sidebar's interaction section: the *recent changes* page. This tension is best summarized in the editorial mechanism called *processes*. According to Ayers *et al.* (2008) 'a Wikipedia processes is a page, or a suite of pages, normally found in the Wikipedia namespace where editors discuss proposed decisions'. As such, 'processes are public, open, and transparent' but as an act of contribution they are individual and private (Ayers *et al.*, 2008: 217). This conflict is most rigorously displayed in, what Ayers *et al.* (2008: 392–393) call, a *single-purpose account*: 'when an editor creates an account and then only edits one article (or a small group of related articles), he or she may be some kind of activist with a narrow focus, rather than an encyclopedist who views the Wikipedia project in the larger sense'.

These different mechanisms show that editing is not only transparent but also permanent, as Wikipedia keeps a complete record of all changes that are made to any page (the page history is viewable by everyone). Each time a page is changed, a new version is saved, but the old versions are not lost (Ayers *et al.*, 2008: xxi). The necessity for management of online communities cannot be underestimated, as Petric and Petrovcic (2014) explain, since the misconduct of community members in the form of hate speech, trolling, cyberbullying and deception can have damaging effects on the quality of the information and social processes in an online community. But the effect of rules goes further: they also attempt to make its group identity and social cohesion more transparent. This is evident in the case of Yiddish Wikipedia as it provides accountability for the long-lasting historic argument laid out earlier.

The transparency and codification (manifestation of the meta-discourse in the Wiki history page) imposed by Yiddish Wikipedia reflect the intricacies of Yiddish's chronoscopic battle with Modern Hebrew. The anonymity in a hybrid community, such as Yiddish Wikipedia, on the one hand, facilitates a more uninhibited conversation while the familiar environment defined by membership in the Yiddish (scholarly) community, on the other hand, has a tendency to pull the discourse into the realm of Yiddish's history and current identity politics. We arrive at a certain amount of predictability in an unpredictable environment. In order to comprehend this interwovenness, Bruno Latour's actor-network theory, Silverstein's agency and local community theory, Sköld's documentary practices theory or Fishman's concept of ethnic identity come to mind. Though coming from different disciplines (sociology, linguistics, informational science), these scholars share the understanding that the web community as a community of practice is foremost of a social nature in which identity practices are instrumental. This proposition will be examined in a proposal to shut down the Yiddish Wikipedia, which was brought forth in 2007 and ultimately rejected shortly afterward. The meta-discussion of the proposal reveals how, ironically, the lack of trust that is so necessary for virtual communities and conferred presumptively has been harmful for the existence of the Yiddish Wikipedia site, but at the same time, *mutatis mutandis*, has made the existence of Yiddish Wikipedia so important.

It shows, as a matter of fact, how Yiddish Wikipedia represents an uncommon case of a virtual community.

'Be wary of the most contentious topics, such as religion and politics [...]'⁶ or Yiddish Wikipedia as a Site of Resistance

Yiddish's present in the past

Yiddish is no stranger to the internet world and Ross Perlin (2014) thoroughly explicates the type of online representations that Yiddish has assumed, such as Yiddish online journals, tweets, blogs, message boards, newsletters, accessible broadcasts, online book clubs and book centers. As a diaspora language, the Yiddish-speaking community sees possibly even more of an advantage to the virtual venue: it is fast and efficient as it connects Yiddish speakers from all over the world – a quality that is highly important in the Yiddish-speaking world which is divided between the *'frum* [observant] and *fray* [secular] Yiddish speakers' who have limited contact with each other. Despite the comparatively small number of Yiddish Wikipedia users (in 2006, Berger estimated that on average 100 visitors used the Yiddish Wikipedia every month; compared to Hebrew estimated at 10,000 a month), the importance that Yiddish Wikipedia has gained in the community is not to be dismissed easily (Berger, 2006).

Looking at Ostrom's (2000) theory of social norms in collective actions, the one that is, I maintain, most operational in the Yiddish Wikipedia community is the first principle that argues for 'clearly defined community boundaries' determining 'who has rights and privileges within the community'. This is no coincidence considering Yiddish's past. The reoccurring mutual interference between Yiddish and Hebrew language and culture is not a happenstance as can be seen in a meta-discussion about the aforementioned proposal to shut down Yiddish Wikipedia; it reveals how the historical issues between the two languages are intrinsic to the complications inherent to Yiddish Wikipedia.⁷ This discussion is crucial because it discloses the systemic difference between Yiddish Wikipedia and other Wikipedias such as English Wikipedia. I argue that while other Wikipedia entries might display opposed views, the disagreement remains purely factual.⁸ The implicit and explicit arguments embedded in the proposal originate from cultural politics, as I maintain – products of a long-standing ideological and moral war that has been responsible for a hostile discursive field.

The proposal consists of three parts: the argument for closure of the site, the argument against closure and the general discussion. The argument for closure of the site is that there are 'no bureaucrat' or 'trustworthy potential admins' (sic). Most importantly, the charge states that the active members of the site don't know what Wikipedia is all about as each user 'has [his] own agenda' and that 'sysops on the Yiddish Wikipedia are not native speakers of Yiddish' (Yonatanh, 2 June 2007).[9] This proposal was put forward by one participant who allegedly goes by many usernames and who is identified as a Hebrew Wikipedia admin who is pursuing this proposal for personal reasons, as one defender claims ('silenc[ing] a whole community because one guy from another language has a problem with other wikipedians inflating his project', Judl, 2 June 2015). On the other hand, another defender accuses the proponents of the Yiddish Wikipedia shutdown of 'malicious intent' and calls them 'liars' and 'abusers' (Jeo 100, 3 June 2007).[10]

Indeed, in the 'General Discussion' section of the proposal, one of the contributors disclosing himself/herself as 'one of the three admins on loan as "regents" from he-wiki (i.e. Hebrew Wikipedia) to yi-wiki (i.e. Yiddish Wikipedia)' identifies the Yiddish Wikipedia site as 'troubled' with 'many dire problems [...] ridden by horrible relations between warring sides'. A mutual permeation of the two Wiki sites and especially the attempted interference of the Hebrew Wiki on the Yiddish Wiki are repeatedly commented on in the defense part of the proposal. The meta-discussion argues that, increasingly, matters concerning the Yiddish Wikipedia have been openly discussed in the Hebrew Wikipedia accusing the Hebrew Wikipedia of 'abuse of power' in trying to alter the content on the Yiddish Wikipedia; a well-needed separation of the two Wikipedia sites is missing.

The aforementioned elusive concept of 'Judaism' –notions of ethnicity, folk, nation and religion – (even more with the establishment of the Jewish State of Israel that has allowed for a national/secular definition of Jewishness) has led Fishman (1985) to conclude that ethnicity is indeed a key factor in the maintenance of the Jewish identity. According to the eminent scholar, ethnicity can be identified from within, but more properly understood, it represents opposition to outside association and affiliation. Ethnicity is, in effect, a form of negation; it exists as the Other, self-perceived, experienced and emotionally evaluated. Yet, as an internal formation, ethnicity carries a collective sense of kinship, cultural continuity across generations, in which responsibilities and obligations become the links to one's own kind, as a highly conscious outward-oriented ideology and a platform from which to articulate the discourse of difference with the outer community. Boyarin raises the difference between the two concepts 'identity politics' and 'politics of identity' (Boyarin, 1996: 4). While his argument *is* that there is a difference between the two concepts, I maintain that the lack of a distinction is in place in the Yiddish Wiki case. The politics of identity ('a constitutive problematic of everyday life') have become identity politics ('[...] explicitly and consciously organized writing, discourse, or action').

This explains the existence of the two linguistic domains, Hebrew and Yiddish Wikipedia, as such. The mode of discourse, however, needs further explanation. I maintain that while Yiddish and Hebrew were once primarily marked as religious languages, today, both languages have assumed more of an ethnic status signifying a weakened form of religious identity. Whereas religious languages supply a rigid membership in the respective communities, ethnicity is a more fluid concept as social identity enters the discourse and becomes a strong(er) factor in the make up of ethnic identity. Following this thought, ethnic identity has drawn attention to collectivity as it carries a collective sense of kinship, but social identity additionally highlights subjectivity. Lastly, while traditional models of identity are based on the host culture, virtual communities do not factor those in. I argue that all of these changes push the subjective agency into the foreground.[11]

A word needs to be said about the relationship between religious and ethnic identity. Van de Vijver *et al.* (2015: 37) are reluctant to '[reduce] family, religious, and linguistic identity to their ethnic identity', arguing that 'strong linguistic or religious identities do not necessarily coincide with ethnic borders' (de Vijver *et al.*, 2015: 40). Much as their research observes that religious and ethnic identities are coexisting, equally strong variables in instances of cultural diversity, I argue in my work that religious identity can indeed 'subside' in ethnic identity. For this, I argue, we need to differentiate between religious identity that entails religious practice and religious identity that expresses 'merely' religious belongingness. In the case of religious belongingness, religious identity has become less salient as markers of social identity (e.g. religious language becomes less important) have adjoined; the domains of identity have shifted. This thought will be further elaborated in the following section in the discussion about agency.

The Whole Is Other than the Sum of Its Parts

The complications with the Yiddish Wikipedia emerge from the basic issue of whose interests are served by having a Yiddish Wikipedia. According to one contributor on the proposal site, one can distinguish three groups: the first group is the ultra-Orthodox Jews who are 'religiously and ideologically driven and very difficult to work with'. The second group consists of the young educated people who are trying to 'resurrect the language "artificially" and restore Yiddish culture to its past glory' but who are not organized. Last but not least, there are the old people who are not really involved at all.[12] More important than identifying these (historically familiar) three groups is understanding the argument for the continuation of the Yiddish Wikipedia.

The first argument that takes shape in the meta-discussion of the proposal sees the purpose for the Yiddish Wikipedia like any other Wikipedia site: to upload and share factual knowledge as commented in 2007 by one defender that 'most of the 3,500 articles on the Yiddish Wiki are totally encyclopedic'. Another defender calls it, 'an active Wikipedia with thousands of good articles' (Dovi, 2 June 2007). The second argument is different as it alleges a bona fide approach: it is not for the people to share factual knowledge but we owe it to the language and the heritage as one of the defenders maintains, 'the language clearly deserves its own Wikipedia [...] Yiddish is spoken by Three Million people' (Philip J, 5 June 2007). Another contributor argues that 'the whole idea of Wikimedia foundation is not to give the knowledge to those who have it already. Our goal is to give it mostly [...] to the language that needs it' (Judl, 1 June 2007).

This line of argument makes sense especially if Yiddish is seen foremost as an ethnic language rather than as a religious language. Ethnicity carries and communicates tradition, which is always documented through the past in which memory is an operating tool. The discourse of religion, as I argue, is, on the other hand, mostly anchored in the present by means of practice, and occasionally oriented toward the future as it is deistic. Cohesion, then, in the case of Yiddish Wikipedia, is developed not through what *is* but rather, what it should be, informed by what it was. This fluid relation between the past and the present (but also the self and the community that builds ethnic boundaries according to Fishman) – this mode of being is achieved through agency (Fishman, 1985: 47). In Silverstein's (1998) theory of agentive subjectivity, language assumes a more emblematic role as he argues that 'through social action, people participate in semiotic processes that produce their identities, beliefs, and their particular senses of agentive subjectivity'. His theory deems culture to be a virtual and always emergent site in sociohistorical spacetime with respect to the essentialisms of which such agents experience their groupness. Language serves as the central semiotic medium or modality through which agentive subjects experience and articulate those cultural processes. As a matter of fact, as Silverstein (1998: 402) points out, language potentially bears an inherently double relationship to the larger cultural processes of which 'it is both emblematic and enabling'.

Agency or agentive subjectivity reveals itself repeatedly in documentary practices such as Wikipedia entries, according to Sköld (2015). Echoing Bruno Latour (1987), Sköld argues that documentary practices reflect the individual's experience but they are also communal configurations of activities ('shared sociocultural practices of people'). He concludes that as documents create and maintain social groups and not just deliver information, they form *communities of interest*. For this, a mutual dependency between the past and the present becomes operative as documents are viewed as products of its context of creation as well as active agents in this context.

In Sköld's framework, agentive subjectivity is fueled by *interest* and it is perpetuated by *memory* – both subjective traits of individuals that require social space. The author emphasizes the importance of memory, discarding the common assumption that memory is merely a static mapping of the past. Rather, he suggests – resting on previous works by Featherstone (2006) – memory constitutes 'the present by putting understandings of the past to work' (Sköld, 2015: 296). In other words, memories inform present decisions as they contribute to the shaping of opinions and actions. While Sköld focuses more on the impact of the past on the present, this chapter argues that memory assumes longer-lasting agentive power as it seeks to shape the future. Memories form the future as they document the present, and, as Yiddish Wikipedia shows, seek to edit the past to become decisive for the future. In essence, the act of documenting serves ideological purposes to maintain a particular culture and to shape the future of the language and culture – a process that Sköld (2015: 297) calls 'the making of communal memory'.

Wikipedia, and for our purpose Yiddish Wikipedia, represents a particular type of documentary practice that is asynchronous (collaborative) as well as nonsequential (the editing or the deletion of an already existing entry). While traditional encyclopedic documentary practices can usually not be edited and are not traceable post circulation, documentary practices in online social networks such as Wikipedia are omnipresent and remain indicative and constitutive of sociocultural life. This observation coincides with Latour's (2011) rationale which identifies the main difference between pre-digital and online communities to be whether an association between the agent and the information can be manifested. Latour (2011) argues that the issue with the nature of the online community is that the individuals are traceable and as such their actions remain accountable as opposed to dissolving into a *sui generis* social network in which (borrowing from Aristotle) the whole is bigger than its parts rendering a cohesive social theory.

Instead, when it comes to the Yiddish Wikipedia, the whole is other than its parts; each entry/article needs to be separately analyzed to understand the motives for the existence of the Yiddish Wikipedia. If the previously pre-digital, purely data-driven approaches caused a discontinuity – discontinuity between the agent and the data/information – by which 'the individual action disappears mysteriously into a *sue generis* structure', the means of digital communities replaced this discontinuity with a continuity linking the data to the agent (Latour, 2011). Latour (2011) concludes that a network is a 'mode of inquiry' as opposed to a 'representation of'.

We need to stop for a moment and digest how the discussed theories are pertinent to the Yiddish Wikipedia. I maintain that Wikipedia's structural design magnifies Yiddish's long-standing struggle. A hybrid platform with its *ad hoc* random content creations accommodates a hybrid community (i.e. emergent community) such as Yiddish in an unknown environment. Social space is always (re)defined through the content of the discussion which is determined, *ex ante*, by the past. If Yiddish (and Hebrew) are indeed ethnic languages, as I have argued in this chapter, Yiddish Wikipedia then, assuming the mode of a virtual community, lacks a host culture so that ethnic identity can only be established in distinction to a virtual or imagined 'other' which ends up being the 'other' determined by the past. Effectively, the agentive subjectivity that is operational in the documentary practices (i.e. the individualism that is operating in the Wikipedia entries in the form of first-time contributions, responses, editing) communicates subjective interest which is primarily driven by memory. Using Sköld's (2015) notion of memory, the present, documentary entries for the sake of identity formation, only makes sense if understandings of the past are put to work. Talk pages, page history and recent changes pages sometimes impede rather than aid the necessary internal cooperative behavior as this is only an extension of a continuing conversation in the form of open disputes. With its inbuilt traceability that is supposed to provide cohesion, the continuity is not necessarily a continuity of the thematic discourse but rather a continuity of the ongoing cultural war as the discussion of the proposal shows.

This nonsequential and asynchronous documentary practice through the open record access often creates tension between building collaboratively a reference, on the one hand, and creating single work, on the other hand. Lee and Suh (2015) explain the continuity of the cultural war on the Yiddish Wikipedia site psychologically using the power of psychological ownership that they define as a 'cognitive-affective construct, a state in which an individual feels as though an object [...] is "theirs"'. The reward comes in the form of a 'pleasurable emotional state' as there are no physical rewards to be gained in a virtual community (Lee & Suh, 2015: 383). The commitment is the 'desire to remain affiliated' with that community. This psychological ownership, as Lee and Suh (2015) continue, is maintained through three behavioral factors that explain the individual's self-investment.

(1) The *relationship to the object*: 'The targets of ownership can become so deeply rooted within people's self-identity that they can be viewed as an extension of the self' (Lee & Suh, 2015: 383); the more people know the topic, the more they consider the contribution as part of the self. By extension, 'exercising control over an object eventually gives rise to a feeling of ownership of the target'.
(2) *Accountability* whether the expected right to hold others accountable for or the expectations of oneself to be held accountable for their contributions.
(3) *Membership* or what Lee and Suh (2015: 383) call 'belongingness'.

The authors see a correlation between the intimate knowledge of the subject, psychological ownership and their affiliation with the virtual community. The more one possesses information about the object of knowledge, the more one gets involved. In this process, the self becomes attached to the creation or as the authors inform, '[w]hen members have psychological ownership, they will have positive self-assessments of themselves as members of the virtual community' (Lee & Suh, 2015: 385). I argue that the very nexus of the type of documentation practice (i.e. for the purpose of memory maintenance), on the one hand, and the nature of online communities, on the other – in other words, *what* Yiddish Wikipedia is for and *how* it is practiced – puts an exceptional emphasis on the agentive subjectivity, to use Lee's and Suh's theory, on the psychological ownership. The psychological ownership shifts the nature of Wikipedia from being a global network to becoming a local network as Yiddish Wikipedia is ultimately only interested in specific knowledge; it only generates knowledge that is subject to and of interest to the particular cultural processes. Locality of communities of interest (and I include here also language communities such as the Yiddish Wikipedia) is only an analogously produced state in a cultural-ideological order (inclusion and exclusion of; a contrastive consciousness of self–other placement that is part of a cultural project of groupness) (Silverstein, 1998: 405). In other words, locality is only manufactured from within by the community's members as it is characteristically recognized as restrictedness of use, frequently with a concept of a hierarchy of inclusiveness where this local cultural sense appears in the guise of local doctrines of meaning. While Wikipedia is a global product, Yiddish Wikipedia has intensified the desire and maintenance of the participants' local identity; locality becomes an identity-relevant dimension of belonging to a particular group, a self-ascription of having a particular culture.

Conclusion

A participant's choice of how to position himself/herself comes through his/her internal conversations about who the participant is and how the participant relates to the discourse – again, an *ad hoc* and context-sensitive process. We have seen how the local dominant discourse with its own values is heavily determined by past experiences; Yiddish Wikipedia as a virtual community becomes an extension of the diaspora community which is, by definition, a displaced community and contingent on its relationship with the respective contact cultures. Displacement and exposure to alien environments create and foster hybrid communities. As a hybrid community, Yiddish Wikipedia shows how anonymity in hybrid communities can facilitate more ingenuous and uninhibited conversations. But the familiar environment, also referred to as locality (membership in the Yiddish [scholarly] community sharing the same interest and discourse) provides a certain amount of predictability in the social processes as most of the discourse concerns Yiddish's history and current identity politics. Perceptibly, *agency* becomes stronger in the management of particular topics, in the performance of identity construction, policy discourse and interpersonal relationships in the absence of face-to-face conversation and in the presence of framed practices. (Social) space then becomes constitutive and agentive in organizing patterns of communication. This is an understanding of social space that is inherent in the definition of the community of practice: Yiddish Wikipedia strives to be a homogeneous community of practice.

The major aim of the present investigation has been to determine the points of insertion of social-psychological variables into the causal spiral that cause the conflicts on the Yiddish Wikipedia. I have determined that *intergroup* behavior/conflict in the form of perceived ethnic identity, on the one hand, and the *interpersonal* behavior/conflict in the form of agentive subjectivity in a virtual community, on the other hand, are responsible for two types of conflict: the objective (i.e. factual) disagreement that is actually a subjective disagreement, and the explicit conflict that is actually of an implicit nature.[13] Ultimately, any conflict arises due to discriminatory intergroup behavior based on the perceived conflict of objective interests seeking to establish a positively valued distinctiveness for one's own group. However, the nature of the discursive method (i.e. virtual community) shifts intergroup conflict to the interpersonal conflict. Any implicit norms are articulated as explicit norms and interactive content moderations promising transparency (with the help of talk page, page history, etc.) turn into deliberative communications. Opposed group interests in obtaining shared/interrelated doctrines about Jewishness promote competition while superordinate conflicting interests facilitate participation. Yiddish Wikipedia has shown us how conflicting interests develop into overt social conflict. Ultimately, it appears that intergroup competition enhances intragroup morale, cohesiveness and cooperation.

Notes

(1)   See Greenstein and Devereux (2009) for a detailed description of Wikipedia.
(2)   See Soldat-Jaffe (2012) or Katz (2004) for more on this topic.
(3)   'Wikipedia: Neutral Point of View'. See https://en.wikipedia.org/wiki/ Wikipedia:Neutral_point_of_view (accessed 27 June 2015).
(4)   Other principles are '[s]tick to discussing the article, and save self-expression for your own user page, use the talk pages for discussing facts and sources, be brief but not abrupt. Be specific about changes you'd like to see, talk pages have a warehousing function, be civil, and make no personal attacks, avoid the absolute no-nos, don't delete comments, and refactor discussion only as a last resort, don't exclude newcomers, problem users show themselves over time' (Ayers et al., 2008: 339–340).
(5)   On cooperative behavior and social networks, see Fowler and Christakis (2010).
(6)   Ayers et al. (2008: 339).
(7)   'Proposals for closing projects/Closure of Yiddish Wikipedia' (2007). See https:// meta.wikimedia.org/wiki/Proposals_for_closing_projects/Closure_of_Yiddish_ Wikipedia (accessed 18 June 1015).
(8)   To understand what I mean by 'factual disagreement' see Hardy (2007). In his article, Hardy analyzes the Israel–Lebanon war from 2006 and maintains that 'the true surprise in Wikipedia's treatment of the [...] conflict is the number of times the article has been edited'. He reveals that the article 'is of about 1,200 words in length [...] [with] 2,500 words of editorial discussion, much of it debating the

relative merits of various Lebanese and Israeli sources that provided the bulk of the information. [...] There were 183 revisions to the story and 13 sources were cited [with] about 30 hyperlinks. The resources used for revisions were [mostly] news organizations (the BBC, *Reuters*, al-Jazeera, Haaretz, *Washington Post*) or other bodies (IDF, Red Cross)'. The point being that while discussion on the Israel–Lebanon war is tense, it does not question anybody's integrity.

(9)   'Proposals for closing projects/Closure of Yiddish Wikipedia'.

(10) The proposal was put forward by one participant and supported by an unregistered second participant whose profile was ultimately shut down on 20 April 2015. 'Creating User:זזלא'. See https://meta.wikimedia.org/w/index.php?title=User:%D7%90%D7%9C%D7%96%D7%95%D7%96&action=edit&redlink=1 (accessed June 2015).

(11) For an excellent discussion on ethnic identity, see Van de Vijver *et al.* (2015).

(12) Harel (June 2007). See https://meta.wikimedia.org/wiki/Proposals_for_closing_projects/Closure_of_Yiddish_Wikipedia (accessed 26 June 2015).

(13) Implicit conflicts are those that can be shown to exist despite the absence of an explicit context.

References

Ahuja, M.K. and Carley, K.M. (1999) Network structure in virtual organizations. *Organization Science* 10 (6), 741–757.

Ayers, P., Matthews, C. and Yates, B. (2008) *How Wikipedia Works and How You can Be a Part of It*. San Francisco, CA: No Starch Press.

Baker, N. (2008) The charms of Wikipedia. *The New York Review of Books* 55 (4), 1–10.

Berger, S. (2006) An entsiclopediye fun stam mentshn. *Forverts*. See yiddish2.forward.com/archive/forverts/2006/1027/item4.html (accessed July 2015).

Bhabha, H. (1994) *The Location of Culture*. New York: Routledge.

Blommaert, J., Collins, J. and Slembrouck, S. (2005) Spaces of multilingualism. *Language & Communication* (25), 197–216.

Boyarin, J. (1996) *Thinking in Jewish*. Chicago, IL: University of Chicago Press.

Brooks, F. (1975) Aristocracy, democracy, and system design. In *The Mythical Man Month: Essays on Software Engineering* (1st edn; Chapter 4). Reading, MA: Addison-Wesley.

Bushera, H. and James, N. (2015) In pursuit of ethical research: Studying hybrid communities using online and face-to-face communications. *Educational Research and Evaluation: An International Journal on Theory and Practice* 21 (2), 168–181.

Featherstone, M. (2006) Archive. *Theory, Culture & Society* 23 (2–3), 591–596.

Fishman, J.A. (1985) *The Rise and Fall of the Ethnic Revival: Perspectives on Language and Ethnicity*. Berlin and New York: Mouton.

Ford, H. (2015) Infoboxes and cleanup tags: Artifacts of Wikipedia newsmaking. *Journalism* 16 (1), 79–98.

Forte, A., Larco, V. and Bruckman, A. (2009) Decentralization in Wikipedia governance. *Journal of Management Information Systems* 26 (1), 49–72.

Fowler, J.H. and Christakis, N.A. (2010) Cooperative behavior cascades in human social networks. *Proceedings of the National Academy of Sciences of the United States of America* 107 (12). See www.pnas.org/content/107/12/5334.long (accessed June 2015).

Greenstein, S. and Devereux, M. (2009) Wikipedia in the spotlight. *Kellogg School of Management Case*. See doi.org/10.1108/case.kellogg.2016.000411.

Grice, H.P. (1989) *Studies in the Way of Words*. Harvard: Harvard University Press.

Hardy, M. (2007) Wiki goes to war. *Australian Quarterly* 79 (4), 17–22.

Katz, D. (2004) *Words on Fire. The Unfinished Story of Yiddish*. New York: Basic Books.

Kumar, S.R., Raghavan, P., Rajagopalan, S. and Tomkins, A. (1999) Trawling the web for emerging cyber-communities. *Computer Networks* 31 (11), 1481–1493.

Latour, B. (1987) *Science in Action: How to Follow Scientists and Engineers Through Society.* Cambridge, MA: Harvard University Press.

Latour, B. (2011) Networks, societies, spheres: Reflections of an actor-network theorist. *International Journal of Communication* 5, 796–810.

Lee, J. and Suh, A. (2015) How do virtual community members develop psychological ownership and what are the effects of psychological ownership in virtual communities? *Computers in Human Behavior* 45, 382–391.

Ostrom, E. (2000) Collective action and the evolution of social norms. *Journal of Economic Perspectives* 14 (3), 137–158.

Perlin, R. (2014) Blitspostn, vebzaytlekh, veblogs: The rise of Yiddish online. *Slate.* February 2014. See: www.slate.com/blogs/lexicon_valley/2014/02/27/yiddish_language_the_mame_loshen_has_network_of_adherents_on_the_internet.html (accessed July 2015).

Petric, G. and Petrovcic, A. (2014) Elements of the management of norms and their effects on the sense of virtual community. *Online Information Review* 38 (3), 436–454.

Rosenzweig, R. (2006) Can history be open source? Wikipedia and the future of the past. *The Journal of American History* 93 (1), 117–146.

Silverstein, M. (1998) Contemporary transformations of local linguistic communities. *Annual Review of Anthropology* 27, 401–426.

Sköld, O. (2015) Documenting virtual world cultures. Memory-making and documentary practices in the City of Heroes community. *Journal of Documentation* 71 (2), 294–316.

Soldat-Jaffe, T. (2012) *Twenty-First Century Yiddishism. Language, Identity, and the New Yiddish Studies.* Brighton: Sussex Academic Press.

Van de Vijver, Fons, J.R., Blommaert, J., Gkoumasi, G. and Stogianni, M. (2015) On the need to broaden the concept of ethnic identity. *International Journal of Intercultural Relations* (46), 36–46.

Wales, J. (2002) 'Jimbo Wales'. See http://en.wikipedia.org/wiki/User:Jimbo_Wales (accessed 7 October 2015).

Wenger, E. and Snyder, W.M. (2000) Communities of practice: The organizational frontier. *Harvard Business Review* 78, 139–145.

Wenger, E., McDermott, R. and Snyder, W.M. (2002) *Cultivating Communities of Practice: A Guide to Managing Knowledge.* Boston, MA: Harvard Business School Press.

Zhitlovski, C. (1904) Dos Yidishe Folk Un De Yidishe Shprakh. In J. Mark (ed.) *Geklibene Verk.* New York: Futuro/Cyco-Bicher Farlag (1955).

Part 3
Faith, Language and Online Televangelism

Part 3

Faith, Language and Online Televangelism

7 Digital Evangelism: Varieties of English in Unexpected Places

Tope Omoniyi

Churches and convention fields are the traditional hallowed grounds in Christendom. In this chapter, we turn to religious ritual via digital technology to explore the issues that it throws up for Sociology of Language and Religion (SLR) theory and practice. Believers Love World Ministry (BLWM) is registered as a Christian charity in Nigeria but its congregants are spread across the world. Its physical headquarters are located at 23–26 Kudirat Abiola Way, Oregun, Lagos, Nigeria. Its virtual home is on YOOKOS (acronym for *You Own the Kosmos*), an internet service provider (ISP). Evangelism is a site of engagement in which the value of codes has been modified as a consequence of unequal access and mobility. I shall explore the various segments of the Global Communion Sunday (GCS) service which takes place on the first Sunday of every month with a focus on the new mobilities enabled and audiences created for some varieties of Nigerian English in unexpected places (Heller, 2011; Pennycook, 2012). We shall consider the transformations that facilitate or hinder such mobility.

Introduction

One of the characteristics of 20th- and early 21st-century sociolinguistics is the seemingly clear delineation of social spaces and the language practices associated with them. This is the framework within which concepts such as speech community and situational contexts were conceptualised. Similarly, the World Englishes paradigm associated specific regions with specific varieties of English such that English as a native language, English as a second language (ESL) and English as a foreign language (EFL) contexts were clearly territorially differentiated and demarcated. However, in the last two decades, both of these disciplinary paradigms have experienced shifts as a result of extensive mobility, not only of people facilitated by improved transport systems that enable employees to commute, for instance, between a home in London and a workplace in Brussels, but also technology-enhanced non-physical mobility that enables the carrying out of management meetings involving participants from multiple corporate locations across the world. The consequence of these mobilities is extensive contact between speakers of different language varieties and thus, the transgression of language-varying boundaries and the destabilisation of territory-based theoretical paradigms. Although my illustrations in this chapter will be drawn from Christian faith practice, similar observations can be made across a number of religious traditions that are drawing on the same digital distributional resources. Islam or, for that matter, any other religion, is no less digital and no less transnational than Christianity.

Data and structure

The data on which the discussions are based derive from four sources, and these more or less frame the structure of the chapter. The data comprise transcripts of online video conversations/discussions from the web page of Christ Embassy (CE), notes and observations from the GCS (June 2015), New Year's Eve Service (December 31, 2015) and the Pastor Chris Digital Library (PCDL).[1] For ease of navigation, the discussions are structured as follows: first, I introduce some of the theories that frame the discussion of digital evangelism (DE). This is followed by a historical background of the broader practice of televangelism. The bulk of the chapter is devoted to themed analyses of data collected from two ministry activities: the Healing School and the Innercity Mission; events of the monthly Believers Love World Ministry (BLWM) GCS; and the New Year's Eve Service with their multifragmented congregation in varied global locations accessing the same varieties of English via digital broadcast. The discussion concludes with a brief section on the PCDL.

Theoretical framing

The frontiers of sociolinguistic scholarship have expanded steadily since Coupland's (2003) call to engage more critically with social theory. Mobility and contact, whether negotiated or forced, by their very nature must entail to varying degrees revisions to the existing social order. With that in mind, interest in the ways in which developments in technology impact language practices across all domains has increasingly attracted the attention of scholars. Deumert (2014) posed the following questions in the introduction to her monograph, *Sociolinguistics and Mobile Communication*:

> What does it mean to be able to speak or write to anyone, anywhere, 24/7/365, and get an immediate response? And what does the current profusion of these technologies mean for the study of language in social life? Do we need to develop new approaches, methodologies and theories? (Deumert, 2014)

In the sections that follow, I shall explore some of the theories that these questions underpin with a critical look at the operations of BLWM.

Domains

The characterisation of social contexts in sociolinguistic theory underlies conceptualisations of domain analysis (Fereidoni, 2010; Fishman, 1971, 1972). The framework enables us to construe the church as part of the paraphernalia of a worship domain alongside other totems such as the Bible, altar, priest, congregation, presbytery, parish and so on. In relation to the development of digital technology, arguably the mode of access to salvation has also been broadened and the ideas of the worship domain and the church as a social space in its core have altered dramatically with transnational delivery to and participation by dispersed congregations organised from a spiritual headquarters. One fundamental issue that is raised by this development is the sacralisation of television and the internet, media that were traditionally associated with secular entertainment and/ or education. The significance of this lies in the potential and capacity of these resources to grant mobility to languages and language varieties.

Language varieties in unexpected places

In the emerging literature of the new and growing field of the SLR, one subject that needs to be addressed is the arrival of missionaries from the ideological South in northern churches, a reversal of the tradition of northern missionaries proselytising in the nations of the South. Pope Francis from Argentina and the Archbishop of York, John Setamu (UK) are examples of this trend in the last two decades or so. One of the interesting consequences of the phenomenon is the emergence of officiating ministers who are non-native speakers of the religious language in predominantly native-language churches. The relevance of this to our current purpose is that whereas the physical movement of missionaries remains subject to the immigration laws of nation states, the digital platform drastically cuts down on the nation state capacity to hamper the mobility of language varieties.

For the purpose of exploring varieties of English turning up in unexpected places as a consequence of evangelism, it is necessary to focus on a religious mission operating out of a non-native English centre. This allows one to tie 'the idea of unexpectedness with the idea of transgression' (Pennycook, 2012: 21; 2007). In Pennycook's (2012: 17) and Heller's (2007: 343) conceptualisation, boundaries and expectations supposedly frame convention and normativity in theory. In reality, the fixity of places and the practices that define them is increasingly being questioned. Both scholars suggest that boundaries are transgressed and consequently language practices are found in unexpected places. Transgression in this study lies in the reversal of the flow in the itinerary of religious missions. The southern converted have become converters in northern contexts, rekindling the flame where religious apathy has gained ground (see Omoniyi, 2012).

Furthermore, the normative one-to-one link between religious faiths and languages and/or language varieties of necessity comes under closer scrutiny and we need to begin to ask if indeed any religion in our contemporary globalised world is confined to any one language or variety of a language. Is salvation more readily accessible in one language/variety than another? Similarly, the architecture and assignation of worship spaces has altered from the early 20th-century notions of the village or town church to today's mega churches and the momentary sacralisation of otherwise secular spaces such as stadia (e.g. Wembley Stadium) and recreation grounds (e.g. the O2 Arena in London) for worship purposes. Even more interesting is the emergence of multiple fractured audience-congregations enabled by digital technology.

We have evidence of the ways in which religion is serving the reclamation of dying languages on the one hand and the manner in which language vitality facilitates the spread of some religions. With the focus of this volume on digital technology and faith practices, my objective in this chapter is to explore technology-dependent religious ritual and discuss what this means for the study of language in faith practice as an aspect of social life. Specifically, my focus is on the mobility of non-native varieties of English that causes them to surface in unexpected places.

Background

Televangelism is not a new phenomenon, having existed for nearly half a century now with Christian missionary activity by American Evangelist Billy Graham becoming a household name in the Christian world by the 1980s. Some of the Billy Graham classics have entered the digital age and are available today as YouTube videos; the Christmas 1983 Web Exclusive is an example of this.[2] From the point of view of language in faith practice, these videos perpetuate the idea of Christianity as an imported religion telecast to postcolonial societies in a homogeneous medium, Standard Native English, in this case American. The choice of the standard variety is understandable because Evangelist Billy Graham's initial target audience had been conservative middle-class America, before broadening to pursue social and racial integration and facilitate the late Reverend Martin Luther King.[3]

What is digital evangelism?

DE is a form of mediatised religious practice, the packaging and broadcasting of religious ritual and the propagation of doctrine via digital technology. DE is not peculiar to any one religion as several religious traditions have discovered and explored the efficiency of digital platforms and have used them for all kinds of religious activities. Religious organisations may explore existing secular platforms such as Facebook, Yahoo, Google, YouTube, Twitter, Flickr and Instagram. Events may be domiciled on a digital platform but may not necessarily be digitalised. The Christmas 1983 Billy Graham Web Exclusive referred to earlier exemplifies this. Secular platforms are exploited for religious functions, though sacralisation or religious organisations may design and launch their own, like CE's YOOKOS and CEFLIX. I shall return to this later.

Figure 7.1 is a screenshot of Pastor Chris Online set up for the New Year's Eve Service (December 31, 2015). It remains open to those who wish to relive the experience of the night that ushered in 2016, tagged 'The Year of Spreading' by the ministry. Figure 7.2 is a screenshot of the front page of Pastor Matthew Ashimolowo's Media Ministries' online presence.

In both of the above cases, the ministers embody and represent their churches. Both ministers are Nigerians; one is based in London and shepherds a largely African diasporic congregation, and the other is based in and leads his ministry from Lagos with a predominantly local Nigerian membership. The latter has church branches around the world. Pastor Ashimolowo is more distinctively Nigerian in speech while Pastor Oyakhilome approximates what might be described as pitched between an American and a Europhone accent of English (International English, cf. Jenkins, 2000). The spelling convention in the text of Pastor Chris's message (below) to usher in the second quarter of 2016 may be interpreted as indicative of an American leaning:

Figure 7.1 Screenshot of pastorchrisonline.com January 9, 2016

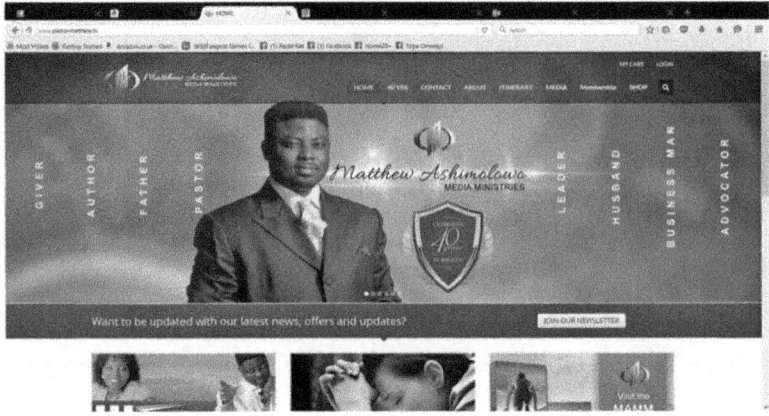

Figure 7.2 Pastor Matthew Ashimolowo of Kingsway International Christian Centre, London

> ..Today, the 1st of April marks the beginning of a new quarter and there's so much to be thankful to the Lord for. Thousands from around the world have been ushered into God's Kingdom through our ministry materials and *programs*, outreaches, churches, partnership organizations and more... (author emphasis)

This could also be the reflection of a more global trend of rising American English variety dominance inspired by Disney rather than the personal preference of the pastor. In spite of this observation, nevertheless, on digital media platforms both pastors' idiolects of English are probably equally as accessible to people from varying language backgrounds.

Spatially contained faith practice allows us to distinguish hallowed from secular ground. This is complemented by temporal delineation in media practice such that religious broadcasts are framed differently from secular entertainment programme schedules. The BBC's 'Songs of Praise' and Channel 4's 'Call to Prayer' at sunrise during Ramadan (2016)[4] represent solemn periods in the broadcast regime with characteristic formulaic ritual language use. We must be cautious though and acknowledge that the sacred and the secular are not always as clearly delineated as we find in the doctrinal and entertainment functions of holy hip-hop (Omoniyi, 2010).

Ministry and online multilingualism

Our data afford us the opportunity to take a critical look at the relative presence of languages online. Although English is the *de facto* official language of the BLWM, some of its resources and events are available in translation around the world. The daily devotional, 'Rhapsody of Reality', also called 'Messenger Angel' by devotees, has been translated into 660 languages and is accessible online too. Pastor Chris remarked in his GCS message (January 3, 2016) that

> In 2015 we distributed 182 million copies of Rhapsody of realities around the world, we set a goal of 120 million we surpassed it. That also did move the message forward in an extraordinary way. Rhapsody of realities is the most circulated book in the world.

Since then, it has become available in a few more languages as the screenshot in Figure 7.3 indicates.

Local faith events evidently take place in many more languages in faith communities than they do online. In other words, online multilingualism richly illustrates what is defined in the sociology of language as societal multilingualism (Grosjean, 1982). The church buildings and convention fields that are considered traditional hallowed grounds in Christendom form part of the contexts of conventional multilingualism.

The BLWM, also known as the Love World Ministry and CE, is registered as a Christian charity in the United Kingdom. Although congregants are spread across the world, its physical headquarters are located at 23–26 Kudirat Abiola Way, Oregun, Lagos, Nigeria. Its virtual home is on a number of platforms: internet service providers (ISPs) including CEFLIX (Christ Embassy Flix) and YOOKOS (You Own the Kosmos).[5] These physical and virtual realities are neither exclusive nor discontinuous. In fact, their activities are seamlessly interconnected such that the borderland between them is barely noticeable and remains largely uninvestigated (see Omoniyi, 2013). However, the focus in this chapter is on multiple audience fracture

Figure 7.3 Rhapsody of Realities

and multiple contact zones, prospects of faith mobility and the appearance of varieties of English in unexpected places (Heller, 2011; Pennycook, 2012), varieties that congregants may otherwise never encounter in the context of a conventional church service.

Digital props

The idea of digital props here is an allusion to the component of Schank and Abelson's (1977) Script Theory; the materials necessary for the acting out of a script, in this case a contemporary evangelical script. The use of social media resources as digital props for the purpose of evangelism facilitates sharing religious doctrine in online communities. But this also means that evangelism is not territorially confined. However, some religions have spread to regions where limited applied technology and, less so, digital technology have registered any kind of presence. Thus, the local language practices may not reflect the reality of the metropolitan voice at Faith HQ. This experience is not necessarily a perspective exclusive to Christianity. A case can be made therefore that digital props may serve both sacred and secular purposes. Increasingly, faith leaders access electronic versions of the Holy Book via an application (app.) on a tablet for sermons and biblical studies.

Staying on the issue of social media digital props, Facebook as a platform has been explored for various BLWM's activities. In Figure 7.4, we find the online domicile of a Romanian branch of the Church in Bucharest.

In the screenshot, the accompanying text is in Romanian with a translation tool at the end of it that enables one to generate the text in English. Translation programme props may be resourced with particular varieties of a language to serve a specific religious community, in which case, the unexpected in the source language or variety may be managed in translation. In other words, if a preacher preaches in a variety of English

Figure 7.4 Screenshot of Christ Embassy Bucureşti's Facebook page. December 31, 2015

that would not normally be available in the Romanian media, say Nigerian English, the translation or subtitling can subvert the arrival of the latter in the Romanian linguistic landscape. What has not been investigated but remains plausible is the fact that faith members located in regions where the religious variety of language is absent may be motivated to acquire it by the desire to access information directly from the Man of God rather than in translation. Thus, through the religious order, as we find also with popular culture (see Omoniyi, 2015), a language or language variety may spread abroad.

CE ministry's presence online has also grown tremendously as Pastor Chris Oyakhilome's vision has spread to different parts of the world and internet technology has been engaged as a strategy to facilitate the development of a global CE family. The two ministry-owned platforms, CEFLIX and YOOKOS (Figure 7.5), constitute hubs for a community of devotees.

Programmes such as the InnerCity Missions for Children (Figure 7.6) key into secular issues that are core concerns of international organisations like the United Nations, UNICEF and UNESCO and depicts the church as a social institution. Other church-sponsored activities like the Healing

Figure 7.5 YOOKOS

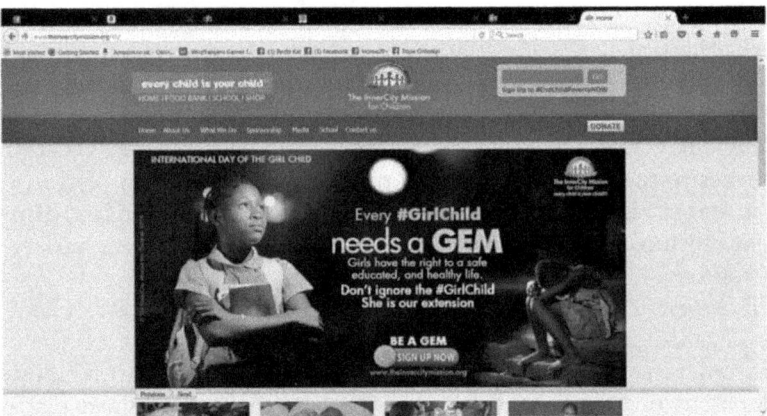

Figure 7.6 Screenshot of CE's InnerCity Mission. January 9, 2016

School (Figure 7.6) have both physical and digital participation. In these instances, what is of interest is the fact that digital media not only facilitate their effective operation, but more significantly mobilise several varieties of English as vehicles of access to evangelism in unexpected places. These are varieties spoken by the pastors involved in the faith healing process and the varieties spoken by the patients who are drawn from different regions of the world into the same social space, the venue of the Healing School. The fact that a Nigerian pastor is providing spiritual and missionary leadership for populations that first exported Christianity to Nigerians transgresses the conventional trajectory of doctrinology.

The Healing School

The Healing School operates three sessions yearly. The autumn session is held in Johannesburg, South Africa. The summer session is held in Ontario, Canada, and the June session is held in Lagos, Nigeria. All three schools are located in officially English-speaking countries that fall into two different circles on the Kachru (1986) model; Canada and South Africa lie in the Inner Circle while Nigeria belongs to the Outer Circle. Still, Canada and South Africa differ significantly in their linguistic ecology and are likely to respond to other circle varieties differently. All three Healing School sessions admit patients from all over the world. On the website (enterthehealingschool.org), a drop-down option bar allows one to select from four languages of access: English, Spanish, French and Portuguese (Figure 7.7).

The interesting point about the drop-down option bar is that the four languages it makes available are the four ex-colonial languages which are the *de facto* official languages of sub-Sahara African countries. This,

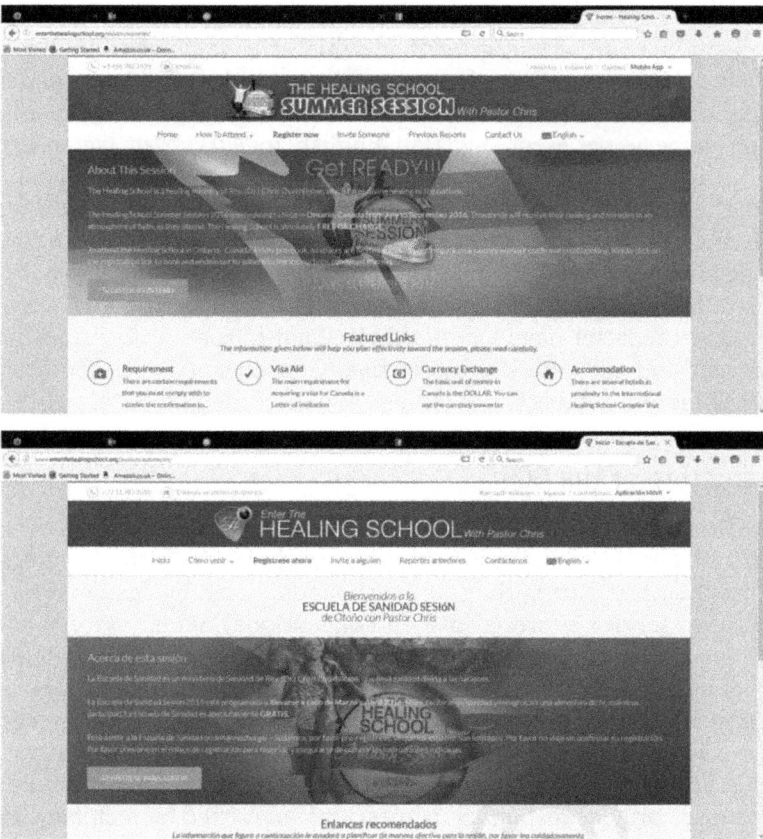

Figure 7.7 (a) The Healing School – Summer Session 2016. (b) The Healing School – Summer Session (Spanish version)

arguably, may be interpreted as a perpetuation of the colonial language legacy. However, the more significant implication is the probable indication that the target audience may be educated and middle class considering that educational instruction takes place in these languages. This is written text, and it might be worth exploring if there are any indications that these are second language varieties too.

The Global Communion Service

Online activities that I describe as digital faith practice include the GCS on the first Sunday of every month with churches in Europe, North America (USA and Canada), the Middle East, Africa and Asia joining in via digital technology. This is a huge departure from the traditional practice of situated worship which had congregations with predictable language experience. This marks a shift in the dynamics of the delivery of salvation. Evangelism is a site of engagement in which the value of codes has been modified as a consequence of unequal access and mobility. In exploring multiple audience fracture, we garner prospects of faith mobility enabled through some varieties of Nigerian English in *unexpected places* (Heller, 2011; Pennycook, 2012). Transnational participation on the digital platform implies that varieties are conveyed by more than one voice. In other words, it is possible, for instance, to encounter a variety associated with a country that one may not have visited. In this regard, the impact of DE may be said to be similar to that of traditional mass media such as film and television except that the latter have a mono-directional format largely from sender to recipient and they are less interactive. Consequently, the contact zone established is less complex.

The structure of the GCS

At CE, the first Sunday of the month, as I pointed out earlier, is observed as a day of fasting globally. The service concludes with the congregation breaking their fast during the communion conducted by Pastor Chris. The communion service is preceded by a conventional Sunday service during which programmes are locally officiated in each of the participating churches by the resident pastors.

Structurally, the GCS has evolved over time to its present three-segment format: Question and Answer, Announcements and Communion. The question and answer segment includes contributions by senior pastors who are invited to provide responses to questions submitted by members. Pastor Chris usually summarises these responses or critiques them as appropriate. There was, for a season, a testimony segment that was coordinated by one of the senior pastors and a selection of testimonies taken from brethren around the world was scheduled monthly into the Communion Service.

A number of things are worth commenting on in Figure 7.8.

(1) The GCS is a live as well as an online event from 4pm (GMT).
(2) It is on a Sunday (but most evangelical ministries now have programmes on other days of the week besides Sunday – these include midweek service on Wednesdays, last Friday of the month All Night service, Saturday Community Evangelism).
(3) There is an open invitation whose addressees are not specified.
(4) During the GCS, Pastor Chris performs mixed roles. He is a religious media host in the television frame, but he is more a spiritual leader (Man of God) in the communion service frame. In these roles, the settings and props including his clothes are different; he is in a suit in the former role (Figure 7.9a) and in a robe (Figure 7.8) in the latter role.

Figure 7.8 GCS June 7, 2015

The first mass congregation test was held during the service on Sunday July 14, 2013. A couple of quick observations are worth noting here. The test was conducted in English. It is not certain how many CE churches in non-English-speaking countries participated in the exercise. It would be interesting too to explore what the attitudes of those congregants were to the idea of the test and more specifically to such a crucial role of English in their experience of church. Figure 7.9b, in a sense, addresses the first of my two observations.

This portal enables the church administration to establish the size of participation per region in study activities. These are often framed in competitive terms so that the most active region gets acknowledged globally, for example, during the GCS. I shall address next the impact of audience fragmentation on our understanding of language varieties in unexpected places modelled on media theory and analysis.

Multifragmentation

One of the evident consequences of DE is congregants' fragmentation (Chandler & Munday, 2011). This has micro and macro dimensions. The micro dimension of fragmentation concerns the split of congregants into main venue and overflow groups. The micro dimension is a phenomenon of single-venue events. For example, when a congregation is larger than the church building can accommodate, an *overflow* is created often using canopies

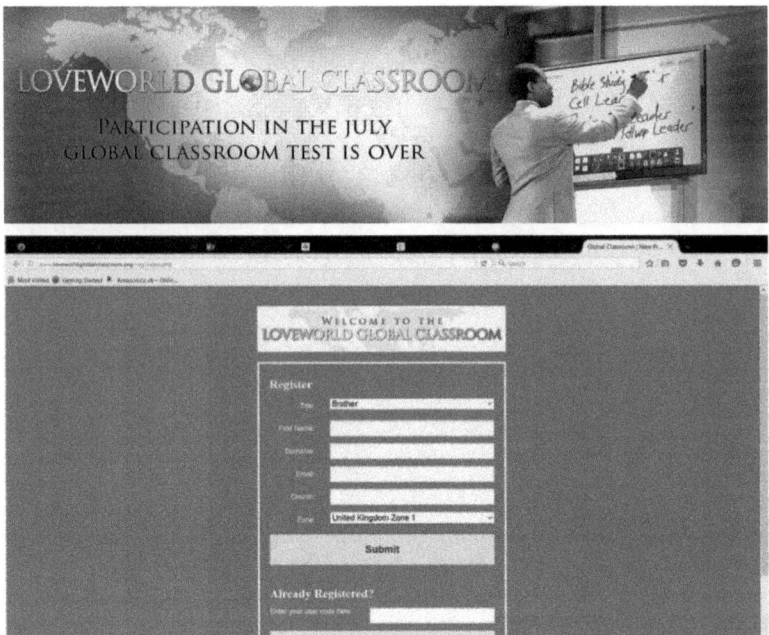

Figure 7.9 (a) Pastor Chris as teacher at the smart board on GCS. (b) LGC Registration Portal

or the immediate outside space generally, which are then served by video link to the activities taking place inside the main arena. The latter congregants may have no real sighting of the pastor unlike their peers in the main hall. In contrast, the macro dimension concerns transnational events coordinated from the administrative core of a religious ministry. In the excerpt below taken from the transcript of CE's online New Year's Eve programme hosted by Pastor Sam, participants from Seoul, South Korea, and the United Arab Emirates contribute to a global talk show. The entire show lasted for 21 minutes and 15 seconds and generated 14 pages of transcript,[6] the entirety of which for space considerations has been abridged and is provided in the Appendix. The abridged version runs from 06:00 to 11:25 and includes the immediate contextual information necessary for full comprehension of the mobilisation of language resources taking place. The interlocutors have not been given pseudonyms since the recording is in the public domain.

Excerpt:
06:54
Christ Embassy invites you to the December 31 service with pastor Chris taking place in Lagos Nigeria and several other centres all over the globe

07:02

pastor Chris: keep this in your heart no matter what you go through, from January to December, you have been triumphant

07:11

we are listening from Seoul, South Korea, we are highly expectant and look forward to what the Lord has for us

07:16

and we are about to celebrate the New Year's Eve with our Man of God pastor Chris

07:21

and all the reverence of the Christ embassy United Emirates is excited to know what is there for this particular year,

there would be rejoicing and thanksgiving for another successful year in this once in a year meeting with pastor Chris

The excerpt opens with an advertisement for the New Year's Eve event with some of the text spoken by Pastor Chris. Using tools of media narrative analysis (Bell, 1998) adapted from Labov's (1972) theory of oral narrative, we are able to establish the orientation of the speech event through the locatives, nominals and time references. In both of the contributions from the two locations at 07:11 (Seoul, South Korea) and 07:21 (United Arab Emirates), congregants/listeners in other parts of the world listen to varieties of English that they may unlikely be exposed to in their local church settings during a regular Sunday service. Similarly, those contributors also listen to the Nigerian variety of English spoken by the pastor/interviewer based at the headquarters in Lagos. Table 7.1 is a compilation of segments of texts from a transcript of the entire interview excerpted above which can be used to illustrate language varieties in motion. A trained and highly competent transcriber prepared the transcript. The items on the left of the table appear in the transcript while those on the right are the speaker's intended texts determined from context. The differences result either solely from various problems in the phonological processes of production in the speaker's variety, including vowel replacement, reduction, lexical substitution, etc., or in conjunction with reception problems due to non-familiarity with the language variety on the part of the transcriber. Whatever the case, we have instances of language experience we can classify as occurring in unexpected places and/or transgression.

It is not certain whether the observed differences between the excerpts from the transcript and the assumed intended alternatives are definite variety features or errors in the individual speaker's repertoire, or indeed instances of mishearing on the part of the transcriber. Whatever the case, it does not really matter. What matters and what is important is that within the framework of the discussion on digitally facilitated mobility of varieties of English, the addressee population who become accustomed

Table 7.1 Language varieties in motion

Transcript	Intended	Probable Process
Reverence	BRETHREN	?? transcriber mishearing
be bald about the word of God	be BOLD about the word of God	[ɔ ʊ] to [ɔː] monophthongisation
that word from a man of God	that word from OUR man of God	[aʊə] to [a] monophthongisation. Nigerian Eng. doesn't have reduced vowels so often as Br Eng., so ˌus' doesn't have schwa
working via breaded	WITH OUR BRETHREN	?? transcriber mishearing
when he gave was that message	when he gave US that message	[əz] to [wɔs] Intrusion or lengthening. Nigerian Eng. doesn't have reduced vowels so often as BrE, so ˌus' doesn't have schwa
but you see she led emphasis	but you see she LAID emphasis	[eɪ] to [ɛ] monophthongisation and shortening relative to BrE
One our man shared with us our message for the year	WHEN our man shared with us our message for the year	[wɛn] to [wan]
and that's why, otherwise, how we would be able to have all of those wonderful testimonies	and that's why, otherwise, how WOULD WE be able to have all of those wonderful testimonies	ˌwould we' to ˌwe would' modal transposition. This is consistent with some EFL varieties.
every part of the word, were going to lay on it	every part of the word, WE'RE going to lay on it	ˌwe're' to ˌwere' by modal contraction+reduction
you know, from it, I do know if there is any other testimony	you know, from it, I DON'T know if there is any other testimony	ˌdon't' to ˌdo' by negator ellision/phonological reduction+vowel lengthening
this sister was one day in our room	this sister was one day in A room	[a] to [aʊə]. Nigerian Eng. vowel lengthening.
completely blind, not seen anything,	completely blind, not SEEING anything,	[siːɪŋ] to [sɪn] by homophony

to the options in the transcript will grow. It is difficult to predict the proliferation of those forms; however, they seem like useful material for parody and humour as we already find in some YouTube parodic videos of various speech-based national identity stereotypes.

The Pastor Chris Digital Library

The PCDL is the final site in this discussion of language practices associated with DE. The portal is a bilingual one available in both English and French. Investigation has not revealed any clear rationale to explain why there are two portals, one in each of these two languages. One explanation may be that the 'local' audience region for CE comprises Anglophone and Francophone sub-Saharan Africa. CE presence in Francophone Africa is understandably relatively smaller than in the Anglophone region. In some Francophone countries, the operation is so small that a virtual zonal pastor is in charge as we find in Togo. In other words, the pastor is virtually rather than territorially situated. In effect then, digital technology is crucial to the sustenance of the branches in Francophone Africa. As far as language is concerned, this forms part of the incursion of English into Francophone Africa and attrition of French in the religious domain.

The PCDL site holds data in various modes: written and spoken. The materials that include Pastor Chris's teachings in videos, audio tapes, e-books and archival records of ministry events are accessible for purchase after registering on the portal (see Figure 7.10). Subscribers purchase credits to enable them to access the resources. This digital resource is relevant to our discussion in this chapter because of the global access it gives to the varieties of English contained in the recordings, thus facilitating their turning up in unexpected places.

It is interesting that the portal above advertises a mix of scriptural materials in English and French but the site template is exclusively in English; the browse options are all in English:

BROWSE
All Messages
Popular Now
Genres
Tags

The PCDL's bilingual resources are a far cry from the 667 languages in which the 'Rhapsody of Reality' exists. However, it may be an indication that the languages are also hierarchised status-wise with English and French at the top. As far as language varieties are concerned, the materials in the PCDL are mediated through the production process and so harmonisation of varieties is a possibility. Ordinarily, the variety of English associated with Francophone Africa is an EFL variety by virtue of the

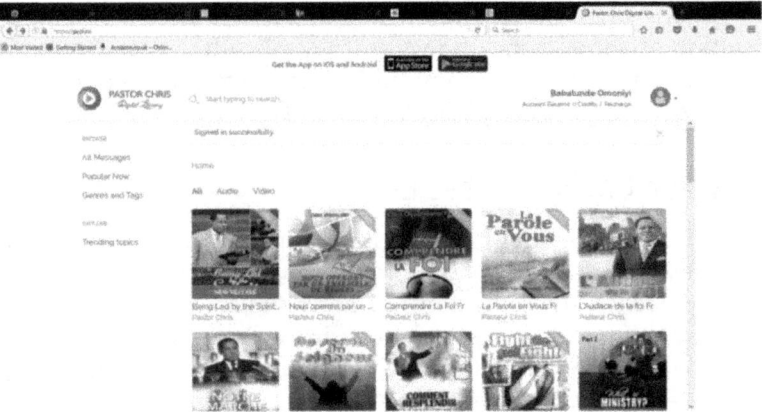

Figure 7.10 PCDL

countries being Expanding Circle countries (Kachru, 1986). Thus, a Nigerian ESL variety accessed on PCDL video by a Togolese church member in Lome would be another instance of Nigerian English turning up in an unexpected place in view of the divergent colonial histories of Nigeria and Togo as British and French colonies and in spite of the geographical proximity and co-membership of the Economic Community of West African States.

Conclusion

In the foregoing sections, I have attempted an exploration of how DE operates to orchestrate the delocalisation of non-native varieties of English. In the old religious order, the varieties that enjoyed extensive mobility were also those associated with colonial political authority, with missionaries moving from Europe to what were to become subsequent colonial footholds. As I pointed out, the movement of missionaries in contemporary times has a different dynamic and has seen ministers from the former colonies evangelising to metropolitan congregations in Europe. In addition, in the 21st century, digital technology has caused a further change in the dynamics of access to salvation through assignment of its resources for the propagation of religion on digital platforms. What has, however, been a significant development from the perspective of our thesis in this chapter is the availability of religious text, in both spoken and written non-native English language varieties, through digital platforms and the concomitant rise in attunement to these varieties in unexpected places.

Notes

(1) Pastor Chris refers to Dr Chris Oyakhilome, spiritual leader of the Believers Love World Ministry.
(2) See http://billygraham.org/video/1983-christmas-special/.

(3)   According to the Billy Graham Evangelistic Association website, Evangelist Billy Graham invited the Reverend Martin Luther King to share the rostrum with him at crusades a number of times in the 1960s and also bailed him out of police custody.
(4)   See http://www.telegraph.co.uk/news/religion/10154170/Channel-4-defends-decision-to-broadcast-Islamic-call-to-prayer.html.
(5)   This in a sense invokes the ideology of prosperity gospel that critics associate with many of the evangelical ministries.
(6)   I acknowledge the kindness of my doctoral research student, Lukasz Daniluk, in assisting with transcribing the data.

References

Bell, A. (1998) The discourse structure of news stories. In A. Bell and P. Garrett (eds) *Approaches to Media Discourse* (pp. 64–104). Oxford: Blackwell.

Chandler, D. and Munday, R. (2011) *A Dictionary of Media and Communication*. Oxford: Oxford University Press.

Deumert, A. (2014) *Sociolinguistics and Mobile Communication*. Edinburgh: Edinburgh University Press.

Fereidoni, J. (2003) A Sociolinguistic Study on Multilingualism: A Domain Analysis Perspective. Proceedings of the 19th Northwest Linguistics Conference, Victoria, BC, Vol. 17, 21–28.

Fishman, J.A. (ed.) (1978) *Advances in the Study of Societal Multilingualism*. The Hague: Mouton.

Grosjean, F. (1982) *Life with Two Languages*. Cambridge, MA: Harvard University Press.

Heller, M. (2011) *Paths to Postnationalism: A Critical Ethnography of Language and Identity*. Oxford: Oxford University Press.

Jenkins, J. (2000) *The Phonology of English as an International Language: New Models, New Norms, New Goals*. Oxford: Oxford University Press.

Kachru, B. (1986) *The Alchemy of English: The Spread, Functions, and Models of Non-native Englishes*. Chicago, IL: University of Illinois Press.

Labov, W. (1972) *Sociolinguistic Patterns*. Philadelphia, PA: University of Pennsylvania Press.

Omoniyi, T. (2006) Societal multilingualism and multifaithism: A sociology of language and religion perspective. In T. Omoniyi and J.A. Fishman (eds) *Explorations in the Sociology of Language and Religion* (pp. 121–140). Amsterdam: John Benjamins.

Omoniyi, T. (2010) Holy hip-hop, language and social change. In T. Omoniyi (ed.) *The Sociology of Language and Religion: Change, Conflict and Accommodation* (pp. 205–233) Basingstoke: Palgrave.

Omoniyi, T. (2012) Multilingualism and religion. In M. Martin-Jones, A. Blackledge and A. Creese (eds) *Routledge Handbook of Multilingualism* (pp. 347–365). London: Routledge.

Omoniyi, T. (2013) Constructing local and global in the e-borderland. In R. Rubdy and L. Alsagoff (eds) *The Global–Local Interface, Language Choice and Hybridity* (pp. 205–224). Bristol: Multilingual Matters.

Omoniyi, T. (2015) The man from Africa is on his way: Styling in Nollywood films. *Journal of Language and Communication Studies* 2 (2) 100–126.

Pennycook, A. (2012) *Language and Mobility: Unexpected Places*. Bristol: Multilingual Matters.

Schank, R. and Abelson, R. (1977) *Scripts, Plans, Goals, and Understanding*. Hillsdale, NJ: Lawrence Erlbaum Associates.

Appendix: Transcript A

31st_December_3.mp4

00:00
It's the eve of another and God through his prophet pastor Chris always gives his children a word of strength and hope to take them through the next 365 days as victors every moment of it

00:16
Christ Embassy invites you to the December 31 service with pastor Chris taking place in Lagos Nigeria and several other centres all over the globe

00:31
pastor Chris: keep this in your heart no matter what you go through, from January to December, you have been triumphant

00:38
we are listening from Seoul, South Korea, we are highly expectant and look forward to what the Lord has for us

00:46
and we are about to celebrate the New Year's Eve with our man of God pastor Chris

00:48
and all the reverenceBRETHREN of the Christ embassy United Emirates is excited to know what is there for this particular year,
there would be rejoicing and thanksgiving for another successful year in this once in a year meeting with pastor Chris

01:13
Christian song

01:20
We invite you today to be a part of this divinely orchestrated programme, so that you too can receive that special word of prophecy from the holy spirit that will launch you grandly into the year 2016 with so much grace and glory

01:39
pastor Chris: make sure that you meditate on God's words, take the word of God seriously into your heart, don't let it go, stay in it, walk in it, be baldBOLD about the word of God, hold it tenaciously, don't let it go and you will be amazed

01:54
at what the Lord will do

01:58

don't miss it for anything, this meeting will also be aired life on your local terrestrial stations, on the Internet and also on the major satellite and cable stations worldwide, for more information about participation at the December 31st service call these numbers or visit the website on your screen, God bless you!

03:22

presenter: hello you are welcome to a special account of our programme, the 31st, global service, with our man of God pastor Chris and before I go I would like to say special thank you to our dear man of God for this opportunity, you know, to talk about our expectations and firstly how the year 2015 has been to us, great testimonies and expectations

03:49

towards the coming year, pastor, I want to say thank you so much Sir for this opportunity, you know the 31st global service of a man of God is a great night, a night full of so much excitement, a night of fellowship with God's spirit, a night of thanksgiving, because, you know, it gives us an opportunity

04:14

to thank God for a great year once again, you know, for all the beautiful things God has done for us in 2015 and it's also of the prophetic entrance, what kind of an entrance are talking about? The entrance that launches us to the new year like we did last year hearing that in 2015 is our year of triumph and we were triumphant in every way and we are looking forward to another night, because the Bible says

04:51

the glory of the latter house shall surpass the former and you know a man of God has been building us up, you know, the excitement is building up, you've got just a few more days to that great night, when millions of people from all over the world, not only in our nation, but all over the world are waiting for that word from aOUR man of God, for the year 2016 and I tell you something I'm excited about it

05:17

and I'm looking forward to it, glory to God, and for those of you out there, I'm sure some of you have got wonderful testimonies from this year, it might give you the opportunity to call him during the show and share with us your testimony, but you see tonight I'm not here alone, I have some wonderful people here with me and together we are going to be talking about, the beautiful year 2015 and our expectations towards the coming year 2016 and with me today I have on my far right the wonderful pastor, pastor Paul OndobosieNDUBUISI (PO) from Christ embassy Lagos zone two, you're welcome pastor Paul

06:02
pastor Paul: thank you Sir the
presenter: and next to him I have also with me from Christ embassy Lagos zone four, that's my zone, Pastor Peijo Akinajali (PA)
PA: good evening pastor
P: you're welcome
PA: thank you Sir
P: we have so much to talk about, but watch this and we will be back, thank you, don't go away

06:30
pastor Chris: Happy New Year! Happy New Year! Happy New Year!
Narrator: it's the end of the year and the beginning of another and God through his Prophet pastor Chris always gives his children a word of strength and hope to take them through the next 365 days
It's the eve of another and God through his prophet pastor Chris always gives his children a word of strength and hope to take them through the next 365 days as victors every moment of it

06:54
Christ Embassy invites you to the December 31 service with pastor Chris taking place in Lagos Nigeria and several other centres all over the globe

07:02
pastor Chris: keep this in your heart no matter what you go through, from January to December, you have been triumphant

07:11
we are listening from Seoul, South Korea, we are highly expectant and look forward to what the Lord has for us

07:16
and we are about to celebrate the New Year's Eve with our man of God pastor Chris

07:21
and all the reverence of the Christ embassy United Emirates is excited to know what is there for this particular year,
there would be rejoicing and thanksgiving for another successful year in this once in a year meeting with pastor Chris

07:38
Christian song

07:54
We invite you today to be a part of this divinely orchestrated programme, so that you too can receive that special word of prophecy from the holy spirit that will launch you grandly into the year 2016 with so much grace and glory

08:11
pastor Chris: make sure that you meditate on God's words, take the word of God seriously into your heart, don't let it go, stay in it, walk in it, be bald about the word of God, hold it tenaciously, don't let it go and you will be amazed if at what the Lord will do

08:30
don't miss it for anything, this meeting will also be aired life on your local terrestrial stations, on the Internet and also on the major satellite and cable stations worldwide, for more information about participation at the December 31st service call these numbers or visit the website on your screen, God bless you!

09:04
P: wow, you're welcome back and like I said I have wonderful people here with me to talk about you know, how the message for the year has impacted them, Pastor Paul, you're welcome once again
PP: thank you

09:20
P: can you tell us how the year 2015, your triumph, you know, has been for you as an individual?
PP: praise God, 1st of all I want to use this opportunity to say very big thank you to my father, my coach, my life director, pastor thank you so much, I love you so much, for bringing me to this platform, I am so grateful Sir, also I need to say thank you to my zonal director pastor Bale Udu Tamos, pastor, thank you Sir, I love you

09:48
PP: pastor, the year 2015 has been a glorious year, I wanna, declare the year 2015 to be the year of triumph and for me personally, it has been a year of triumph, let me just give you one testimony, because I have many in 2014, you know, I run a business, you know in 2014, some years in 2014, the business I faced a lot of challenges, every time when we had the project, we had to look for money everywhere

10:25
PP: we were running helter-skelter trying to get more money, it was difficult, but what pastor declared 2015 to be the year of triumph, there is a scripture that the pastor read, Corinthians 2:14, he said thanks to be on to God, who, which always, always concept us to triumph in Christ and make it a manifest of his work and his knowledge, every place, so those 2 words always in every place, I took a hold of it

8 American Muslim Televangelists as Religious Celebrities: The Changing 'Face' of Religious Discourse

Shaimaa El Naggar

Televangelism or preaching religion via satellite television and YouTube has become an important media phenomenon, *inter alia*, within Muslim minority contexts such as in Europe and the United States (Echchaibi, 2011; Tartoussieh, 2013). Despite the popularity of televangelism, little is known about the discursive strategies that televangelists use to construct their identities and appeal to their audiences, which could give us insights into the nature of this phenomenon and its novelty. The aim of this study is twofold: first, using Critical Discourse Studies (CDS) as a framework, I take a critical approach to the understanding of televangelism. I explore the broader dynamics that have shaped this phenomenon and sustained its existence. I argue that televangelism is a complex phenomenon that can be seen within the contextual prisms of media power, the rise of digital religion and the fragmentation of religious authority in the 21st century. Second, I explore the performance of two case studies of American Muslim televangelists who have achieved popularity in the West and who belong to two different generations, namely Hamza Yusuf (born 1960) and Baba Ali (born 1975). I explore their YouTube sermons and demonstrate that a characteristic feature of the sermons of the two case studies is that they navigate between different discourses and project multifaceted identities. I thus demonstrate the development of televangelism that has embraced many varied styles of linguistic and religious expression. Therefore, in this chapter, I engage with the challenge of keeping up with the growing use of the internet and new media for religious expression (e.g. see Campbell, 2013: 1–8), which is being addressed in this book.

Introduction

In recent years, Muslim televangelism or preaching religion via television and the internet, has emerged as an important media phenomenon (Campbell, 2013; Lotfy, 2009). Televangelists' programmes are broadcast on religious satellite channels such as Umma Channel and Islam Channel. In addition, YouTube is an important platform for the mediation of televangelists' programmes. After being broadcast on television, televangelists' programmes are rebroadcast on YouTube to extend their visibility. Televangelists' programmes can also be posted immediately to YouTube to imagined viewers (see below). A characteristic aspect of televangelism is that televangelists integrate entertainment features in their programmes such as music, sound effects and camera movement techniques. Furthermore, televangelism has become particularly popular among Muslim youth who appear to view televangelism as a new approach to religious messages (Howell, 2008; Wise, 2003). In a similar way to public figures, some televangelists have become media celebrities with hundreds of thousands of fans and followers on YouTube and social media networks (e.g. see Gilliat-Ray [2010] on the American televangelist Hamza Yusuf).

Despite the visibility of televangelists' programmes, they are rarely examined from a discursive point of view. Therefore, little is known about the multimodal and discursive strategies that televangelists use to construct their identities, which could give further insights into understanding the nature of this phenomenon and this new type of religious celebrity.

The aim of this chapter is twofold. First, I provide an explanatory framework for the rise of the phenomenon of televangelism by exploring its characteristic features and the broader dynamics with which they are related. I thus go beyond the perception of televangelism as a novel phenomenon towards raising the question of what might be at stake in televangelism and what might be the wider interactions that have contributed to the shaping of this phenomenon and maintaining its existence. Second, having provided an explanatory framework for the phenomenon of televangelism, I examine two case studies of American Muslim televangelists who have achieved popularity in the West, namely Hamza Yusuf (born 1960) and Baba Ali (born 1975) (e.g. see Echchaibi [2011] on the televangelist Baba Ali and Gilliat-Ray [2010: 162] on the televangelist Hamza Yusuf). I examine the self-presentation strategies of the two case studies. There are similarities and differences between the two case studies that are worth noting. Both Hamza Yusuf and Baba Ali are contemporary Muslim televangelists who grew up in the United States and as such they are both American Muslims. They differ, however, in terms of their style of preaching. Hamza Yusuf usually gives relatively long sermons (about 30 minutes to an hour) addressed to co-present audiences in mosques or Islamic conferences. His sermons are aired on television and/or are posted to YouTube. Unlike Hamza Yusuf, Baba Ali's sermons don't exceed 10 minutes. They are broadcast on YouTube and are aimed at imagined viewers. As I will demonstrate below, the two case studies are different in their self-presentation strategies, for example, in terms of the expression of their authority and the degree of (in)formality. To explore the self-presentation strategies of the two case studies, I employ a synergy of the Discourse Historical Approach (DHA) and the theory of Visual Grammar that emanate from CDS (see below). A salient point that this study will illustrate is how televangelists' sermons have become a multimodal field in which verbal language and image contribute to the process of meaning-making. Furthermore, I argue that a characteristic feature of the performance of the two televangelists is that they play multifaceted roles and construct multifaceted identities. Thus, I trace the development of televangelism as a genre that is open to many varied styles. I aim to engage with the challenging task of investigating the growing use of the internet for religious expression, and the manifestations of religious expression online that are being addressed in this book.

In the following section, I argue that televangelism is a complex phenomenon that can be seen within the contextual prisms of media power (Couldry, 2000, 2003a), the rise of info-tainment (Wodak, 2009) and the fragmentation of religious authority (e.g. see Campbell, 2010).

The Phenomenon of Muslim Televangelism

Historically speaking, the use of television as a medium of religious preaching started in the 1950s in the United States (Bruce, 1990: 29–48) when religious institutions sponsored radio and television programmes for the dissemination of religious information. In the 1980s, however, a new form of religious programme appeared. Fast-paced and entertaining, these programmes comprised a series of segments that included songs, talk shows, interviews, film clips and sermons (Schmidt & Kess, 1986: 36). With telephone counselling and prayer requests, callers were automatically placed on a computerised mailing list through which appeals for donations were sent (Schmidt & Kess, 1986: 36).

Back then, it was unusual for religious figures to integrate songs, interviews and recorded clips into their sermons. These programmes attracted wide audiences and as a result of their popularity, religious figures became – for the first time – media celebrities, on equal grounds to popular stars. Televangelism soon became a phenomenon in Muslim majority contexts. As thousands of audiences in the United States gathered around TV sets to watch televangelists' programmes in the 1980s, audiences in Muslim-majority contexts such as Egypt and Indonesia gathered around TV sets in the late 1990s to watch the televangelists who were – broadly speaking – different in their dress code and style of presentation from formally educated preachers (Lotfy, 2009: 17–32). For example, using colloquial language and an informal style, the televangelist Amr Khaled in Egypt attracted huge audiences as many perceived his programmes as a new approach to religious messages that addressed the everyday life of Muslims, and which focused on individual relationships and goals (e.g. Tartoussieh, 2013; Wise, 2003). For example, in his programme 'Words from the Heart', Khaled draws on the religious narrative of the Prophet's life and his Companions to demonstrate the importance of adopting an upright life in which young people choose good friends, eschew the use of drugs and keep good relations with their parents. In Indonesia, the televangelist Aa Gym attracted millions of viewers to his weekly show (Hoesterey, 2008). In a similar way to Khaled, Aa Gym integrated his religious messages with advice on personal development and business management such as promoting communicative skills and managing stress (Howell, 2008: 56–58).

As has been the case with Muslim-majority contexts such as Egypt and Indonesia, Muslims in minority contexts such as in the United Kingdom and the United States soon produced their own 'home-made' televangelists. British and American Muslim televangelists are featured on religious satellite channels such as Huda TV, Islam Way and Islam Channel UK and on YouTube. They give their sermons in English, primarily addressing Muslims in English-speaking communities.

Many features render televangelism a distinct phenomenon. First, unlike formally educated scholars who wear distinct attire and speak in formal language, televangelists use informal language and relate their sermons to the everyday life of Muslims (e.g. Tartoussieh, 2013; Wise, 2003). For example, the televangelist Hamza Yusuf, whose discourse is examined in this study, instructs his audiences to take the responsibility of reconciling between Muslims and non-Muslims and to denounce terrorism. In his YouTube videos, the televangelist Baba Ali discusses topics related to the everyday life of Muslims in the West, for example, meeting non-Muslims at work, friendship, dating and the misrepresentation of Islam in the mass media. Second, televangelists employ entertainment features in their sermons, for example, sound effects, dramatic elements and camera movement techniques. To illustrate, the televangelist Baba Ali makes use of songs, images and acting in his video blogs (see below). Third, televangelists have digital proficiency as they use a repertoire of new media tools such as websites and social media networks. Both Hamza Yusuf and Baba Ali have online visibility; for example, the televangelist Baba Ali has about 19 million views on his YouTube channel 'umma films'[1] in which he broadcasts his video blogs. Hamza Yusuf has about half a million subscribers to his 'official Facebook page',[2] one of many, and his sermons and recordings are generously rebroadcast on YouTube and on blogs (El Naggar, 2014: 199–201).

The novelty of televangelism (e.g. use of music effects and camera techniques in televangelists' sermons) and its popularity call for a critical approach to this phenomenon which can enable us to understand why televangelism has gained popularity among some Muslim audiences and what broader dynamics it relates to. I argue here that televangelism can be perceived, *inter alia*, within the contextual prisms of media power, the rise of infotainment in contemporary media culture and the fragmentation of religious authority. According to research (e.g. Couldry, 2003a, 2003b; Thompson, 1995), media is symbolically powerful. Couldry (2000, 2003a, 2003b) postulates that media power is sustained through a myth or an assumption that 'there is a centre to the social world and that, in some sense, the media speaks for that centre and that we accept that centre's position in our life as legitimate'. Another important theory that can explain the wide-reaching popularity that some televangelists have gained is 'space–time distanciation' put forth by Giddens (1984). The author contends that one inherent feature of modern societies is the 'fading away of time and shading off of space' (Giddens, 1984: 132), for instance, through the rise of print houses and through the expansion of means of communication. In line with Giddens, Thompson (1995: 21) has argued that mediated communication is 'detached from its context of production both spatially and temporally and re-embedded in new contexts which may be located at different times and places'. Thus, the distanciation of 'space–time' offers an explanation from a theoretical perspective of why some televangelists have achieved a transnational popularity since audiences in different localities can appropriate and relate their messages to their own contexts. For example, a YouTube video by Baba Ali entitled 'Who hijacked my religion?', which has about a million views, is critical of how Western media represents Islam and calls on audiences not to take media stereotypes and the action of some 'misguided followers' as representative of Islam. This video blog appears to be aimed at (non-Muslim) Western audiences. Yet, Muslims in non-Western contexts may relate to it within the context of the global awareness of misrepresentation of Islam after 9/11 (e.g. see El-Nawawy & Khamis, 2009: 4).

Televangelism can also be contextualised in the rise of 'infotainment' or the blurring between information and entertainment in contemporary media culture (e.g. see Wodak, 2009). Kellner (2003) argues that 'info-tainment' has infiltrated media culture from science programmes in which visual effects and a conversational style of content presentation are intertwined (Fairclough, 1995: 3–8) to entertaining politics in which politics is fictionalised (e.g. see Wodak, 2010, 2011) and the domain of politics and entertainment is blurred (e.g. Riegert, 2007; Van Zoonen, 2005).

Televangelism is also a manifestation of the rise of digital religion and the fragmentation of the traditional sources of religious authority (Campbell, 2010). One consequence of the proliferation of digital media technologies is that religious texts (e.g. exegesis and work of jurisprudence) have become accessible to wider constituencies of audiences who are able to (re)interpret religious texts (Robinson, 1993: 245).

Though televangelism has received scholarly interest in recent years (e.g. see Lotfy [2009] and Howell [2008] on televangelism in Egypt and Indonesia, respectively), it has seldom been examined in the West. This may be due to the little academic attention given to Islam in minority contexts vis-à-vis Islam in its 'old' majority settings (Hinnells, 2010: 683). Hence, by examining the performance of two popular televangelists in the West, I aim to contribute to the understanding of current televangelism. This takes me to the next section in which I elaborate on the methodologies and theoretical underpinnings of DHA and the framework of Visual Grammar that represent the theoretical framework of the present study.

Methodologies and Theoretical Considerations

In this study, I employ a synergy of the DHA (Wodak, 2001, 2015) and the theory of Visual Grammar (Van Leeuwen, 1996) to explore the performance of the two televangelists under examination. It is important here to explain the meaning of performance, which will allow a justification of why these two approaches are well suited for the study.

Goffman (1959), a pioneering scholar, put forth the concept of identity as performance. According to the author, life and social reality (i.e. how individuals interact) can be conceived of as a 'theatre in which a show is staged on' (Messinger et al., 1975: 32). Another important scholar who discussed the concept of performance is Butler (1990) who drew on J.L. Austin's speech act theory and the differentiation between constative utterances that describe a situation (e.g. 'it is a sunny day') and performative utterances (e.g. 'I name this ship Normandy') through which an action is performed. The author suggests that– in a similar way to performing actions through uttering verbs – we perform 'identities', for example, gender through 'a stylized set of acts', norms and rituals (Butler, 1990: xv). If identity is performance – as Goffman (1959) and Butler (1990) inform us – then it is important to examine both verbal language and visual features and how these contribute to televangelists' self-representation (e.g. Wodak, 2011, 2015). In this vein, a synergy of DHA and the theory of Visual Grammar enables us to examine the self-representation strategies of the two televangelists (see El Naggar, 2015). I would like to explain what these theories entail. DHA and the framework of Visual Grammar emanate from CDS. One underlying principle of CDS is that 'language is not powerful on its own – it gains power by the use people make of it and by the people who have access to language means and public fora' (Baker et al., 2008: 280). In this way, discourse (language in use) is 'social practice'. For example, it is through the use of discursive strategies such as nomination and predication strategies that particular attitudes are created towards social groups (e.g. immigrants), demarcating the boundaries between in-groups and out-groups (see Fairclough & Wodak, 1997; Wodak, 2011: 2015).

With a view on discourse as 'social practice', important concepts in CDS are 'ideology', 'power' and 'critique' (Wodak & Meyer, 2009: 4–11). Ideology can be defined as 'representations of aspects of the world which contribute to establishing and maintaining relations of power, domination and exploitation' (Fairclough, 2003: 218). Ideology is enacted through linguistic realisations, conveying attitudes and mental representations that shape and/ or reshape social practice. As Wodak and Meyer (2009) put it, ideologies 'hold on to assumptions'; for example, gendered discourse assumes that there are intrinsic differences between men and women. Another key concept in CDS is 'power', which refers to 'an asymmetric relationship among social actors who assume different social positions' (Reisigl & Wodak, 2009: 88). One aim of CDS is to explore how power is legitimised (or de-legitimised) through discourse and how unequal power relations are produced and reproduced (Reisigl & Wodak, 2009: 88–89). CDS is, therefore, aimed at 'critique', through 'produc[ing] and convey[ing] critical knowledge that enables human beings to emancipate themselves from forms of domination through self-reflection' (Wodak & Meyer, 2009: 7).

While DHA shares with CDS its theoretical underpinnings, specific to DHA is the examination of the 'context' of discourse which takes into account four levels (cf. Wodak, 2001: 67):

(1)   The immediate language or text-internal co-text and co-discourse.
(2)   The intertextual and interdiscursive relationship between utterances, texts, genres and discourses.
(3)   The extralinguistic social variables and institutional frames of a specific 'context of situation'.
(4)   The broader sociopolitical and historical context, (in) which (the) discursive practices are embedded and related to.

To illustrate the above, one level of context of discourse is the immediate language. This can entail an examination of the nomination, predication and argumentation strategies used in a specific text, among others (Reisigl & Wodak, 2009: 93–96). The second level of context is the intertextual and interdiscursive relationships between texts/genres. Intertextuality refers to the link to other texts 'through invoking a topic, an event or a main actor' (e.g. Richardson & Wodak, 2009: 46). Interdiscursivity indicates that discourses can be linked to discourses on other topics or sub-topics; for instance, discourse on the environment may contain discourses on finance and health and discourse on exclusion can possibly embed discourses on education and employment (Reisigl & Wodak, 2009: 90–93; Richardson & Wodak, 2009: 46). The third level of context is the extralinguistic social/

institutional context, which refers to all aspects of a situation (in which the discursive event is embedded); for example, the degree of formality, the place and date in which the speech event took place, the recipients of discourse and their social and political background (Blackledge, 2005: 18).

It is worth noting that in DHA – as in CDS – discourse is perceived as a 'semiotic' entity that can include verbal language and visual features (see Reisigl & Wodak, 2009: 89). In this vein, the theory of Visual Grammar (e.g. Kress & Van Leeuwen, 1996) enables us to explore the meanings of visual features such as image, colour and dress code, which are particularly salient to performance. Drawing on systemic functional linguistics, Kress and Van Leeuwen (1996) proposed that images – in a similar way to verbal language – can denote action in what they termed 'visual structure', i.e. 'interpretations of experience and forms of social interaction' in an image. The authors recognised at least two types of 'visual structure'. One visual structure is 'narrative' in which participants (things/persons) are represented as being involved in an action through a line of directionality or vector linking the 'Actor' (i.e. the doer of the action) to the 'Goal' (i.e. the affected participant) as in a picture of a father holding a child. Another type of visual structure is 'analytical' in which participants carry attributes rather than perform an action. A profile photo of a young girl may serve as an example. She will be the carrier of attributes, for example, the colour of her hair, her ponytails and her complexion (Kress & Van Leeuwen, 1996: 45–87; also see El Naggar, 2015).

A salient point in relation to the synergy of the two approaches is that – like DHA – the theory of Visual Grammar is concerned with demystifying ideologies and relations of power in discourse. Kress and van Leeuwen (1996: 67), for example, critique the representation of gender in advertisements in which men are frequently represented as 'the doer of the action' whereas women are represented 'as the faithful admirer of his actions'.

After discussing the theoretical underpinnings of DHA and the framework of Visual Grammar, in the following section I explore the self-representation strategies of the two televangelists. Aspects that I will focus on in my analysis include an examination of the situational context, a multimodal analysis of the sermons of the two televangelists and the hybrid discourses through which they navigate. By exploring the above-mentioned aspects, I will deconstruct the discursive strategies used by these two televangelists, as I turn attention to the importance of examining the visual features besides the verbal language in CDS (see Stöckl [2009: 203] on the little attention given to the analysis of visual language vis-à-vis verbal language in linguistics research). As I will demonstrate below, the two televangelists navigate between many discourses and draw on many multimodal strategies to represent themselves, *inter alia*, as media celebrities.

The Self-Representation Strategies of Hamza Yusuf

I would like to start my analysis of Hamza Yusuf's discourse by giving biographical information about him, which relates to his self-representation. Born in 1960 in the United States and a convert to Islam in 1977, Hamza Yusuf carries the title of an 'Islamic scholar' and 'an intellectual'.[3] According to Esposito and Kalin (2009: 78), Yusuf is 'the Western world's most influential Islamic scholar who has built a huge grassroots following, particularly among young western Muslims'. He established the Zaytuna College in California, which mixes traditional Islamic studies with contemporary Western thought. Its aims are to 'train students in the varied sciences of Islam, while also instilling in them a sophisticated understanding of the intellectual history and culture of the West'.[4] In addition, Yusuf is an advisor to some academic institutions on Islam and some philanthropic organisations; for instance, the Stanford University Program in Islamic Studies and George Russell's 'One Nation' project that promotes pluralism in America.[5]

Lending further support to scholars who have indicated that discourses on Islam in the West are undergoing changes (e.g. Mandaville, 2001, 2007), Hamza Yusuf appears to be one of those who 'critique' contemporary Muslim discourses from within.[6] For instance, one of his published articles argues that denying the Holocaust undermines Islam[7]; thus, he attempts to defy anti-Semitic discourses in some Muslim contexts (e.g. see KhosraviNik [2010] on the anti-Semitic discourses by the former Iranian politician Mahmoud Ahmadinejad).

On another occasion, in 2006, Hamza Yusuf claimed that when some Muslims burnt the flag of Denmark, they ended up doing what they criticised the West for: blaming a whole community for the action of a handful of people.[8]

Hamza Yusuf navigates between different genres and projects multifaceted identities. He represents himself as a preacher, a learned scholar or an intellectual. One discourse genre that Hamza Yusuf draws upon is the admonition or *waaz*, which draws heavily on the use of Quranic recitations and deals with the theme of death and 'the transient nature of human existence' (Swartz, 1999: 44). His YouTube video 'The danger of heedlessness', produced by his recording company Al-Hambra Productions, serves as an example. The sermon starts with a recitation of a Qur'anic verse about heedlessness, i.e. inattentiveness to one's action, followed by an elaboration of the meanings of heedlessness in the Qu'ran. The theme of death is invoked through a personal narrative in which Yusuf describes how he was about to lose his life as he was crossing the street with his children. The situational context plays an important role in framing the genre and Yusuf's self-representation, which is realised through many multi-semiotic resources. The video starts with a distant shot showing co-present audiences sitting on the ground listening to Yusuf in a traditional mosque environment. There is gradual zooming-in on Hamza Yusuf who stands at an arch, wearing semi-traditional garb, i.e. a suit, a turban and a cloak. Using light effect, a halo is drawn around Hamza Yusuf's face in a similar way to revered figures. The persona that Yusuf projects – in this instance – is that of a preacher. Important multimodal features here are posture and gesture. At the start of the sermon, Hamza Yusuf starts by reciting a prayer, rubbing his hands against each other as if he is about to pray. He then puts his fingers across each other, conveying a sense of control and authority (see Figure 8.1).

I would like to elaborate on the role of gestures used by Hamza Yusuf since they appear to be a salient aspect through which he stages his identities. As I will demonstrate below, Yusuf appears to use specific gestures when he projects his identity as a preacher and uses other gestures as he stages his identity as an intellectual. In line with research on gestures (e.g. Goodwin, 2003; McNeill, 1992), it can be argued that Yusuf's gestures perform many functions. He uses 'motor gestures' which are 'simple repetitive rhythmic movements that bear no obvious relation to the semantic content of the accompanying speech' (Krauss et al., 2000). He also uses 'iconic' and 'metaphorical' gestures. The former display meanings typographically such as moving the hands in an upward direction accompanying the phrase 'going up' (see McNeill, 1992: 78). *The latter* are visual representations of abstract ideas and categories, e.g. displaying an empty palm hand may indicate 'presenting a problem' (e.g. see McNeill, 1992: 80). The following example occurs at the beginning of his sermon 'the danger of heedlessness' in which he introduces inattentiveness to one's action as a main topic.

Figure 8.1 Hybrid dress code and posture by Hamza Yusuf

Example 1
[Allaho subhano we ta3la said in the Quran eqtaraba lel nas hesabohom
fa hom fi ɣafla mo3redon[9]]
Hamza Yusuf places the fingers of one hand across the other (see Figure 8.1).
This is how he poses throughout the sermon.
That (0.1) [people's reckonings] are [drawing nearer]
       (1)   (2)
*Gesture (1), metaphoric: Right hand tilts to one side and moves swiftly to the left
enacting the word 'peoples'. (Then, both hands move back to the same posture
where fingers are placed against each other.)*
*Gesture (2), iconic: Hands move forward and both hands roll over one another,
indicating the passing of time.*
[And yet people] are in heedlessness, not thinking about that.
*Gesture, metaphoric: Both hands open, forming a circle, representing 'people'.
(Hands then move on to rest, one over the other.)*

While in the 'danger of heedlessness', Yusuf enacts the persona of the
preacher, on other occasions, Yusuf presents himself as an intellectual,
taking the latter to mean: 'a person of recognized intellectual attainments
who speaks out in the public arena, generally in ways that call established
society or dominant ideologies to account in the name of principle or on
behalf of the oppressed' (Hewitt, 2003: 145). He invokes discourse topics
on politics, creating a hybrid genre of sermon/political rhetoric. His sermon
'Islam and the West', given at a conference in Chicago in 2007 is a case
in point. Hamza Yusuf calls on the Muslim communities to denounce
terrorism, which he argues is a modern phenomenon. The discourse
topics that Yusuf invokes include the contribution of the Muslims to
Islamic civilisation, the unity of the Abrahamic faith (Jews, Christians
and Muslims) and criticism of the US government under George Bush's
administration. The latter includes sub-topics on the abuse of Guantanamo
detainees, the invalid claims of the neoconservatives that the United States
is a Christian country and the importance of addressing the Palestinian
cause. By weaving these topics together, Yusuf assumes the role of not only
a preacher who instructs the 'Muslim community' but also an intellectual
and a public speaker who can be critical of contemporary politics.

Again, Hamza Yusuf's performance is constructed through multimodal
resources. Unlike the semi-traditional garb that he wore during the sermon
'the danger of heedlessness', in the sermon 'Islam and the West', Hamza
Yusuf wears a suit, projecting his identity as a public speaker (Figure 8.2).

In Figure 8.2, Hamza Yusuf stands at a podium wearing a black suit, a
colour associated with authority and modernity; for example, it is the colour of
tuxedos and the uniform of the police (Van Leeuwen, 2011: 2). Furthermore,
unlike his performance in the 'danger of heedlessness', discussed above, in
'Islam and the West', Hamza Yusuf recurrently uses his index finger.

Figure 8.2 Hamza Yusuf wearing a suit in the sermon 'Islam and the West

Example 2

I want to say that (0.1) I feel that the spirit here this year is much higher than last year. So I want to speak to you tonight about a very important topic (0.1). That I feel it is absolutely necessary that I speak about because I would not be true to what I am feeling inside. In Russia, a group of people took hostage of a school filled with children and that resulted in the death of several hundred people including the death of at least several hundred children. This atrocity, unfortunately, was done once again in the name of our religion and I feel that we [as Muslims] are suffering all over because of the acts of a handful of people.

Gesture, deictic: His right hand rises, the tip of his index finger is bent forward, on top of a clenched fist. His hand moves forward, as he utters 'we' and 'as Muslims', apparently pointing to the audience, hence its 'deictic' function.
And we [must in one voice] condemn and completely reject the concept of discriminate violence.
Gesture, motor: He raises his hand; his index finger is pointing forward, on top of a clenched fist

As can be seen above, there is a recurrent use of the index finger, which is, interestingly, a gesture used by politicians in their performances. An obvious example is Barack Obama. This shows the convergence of religious discourse towards political rhetoric in the case of Yusuf (see El Naggar, 2015).

Another salient element in Hamza Yusuf's performances is the camera cutting to distant and close shots of audiences. One function of the distant shots is the creation of a spectacle in which massive audiences are shown attending the speech event, highlighting the celebrity status of Hamza Yusuf. The image from his sermon 'Islam and the West' is a case in point (Figure 8.3). While Hamza Yusuf stands at the podium uttering: 'we must in one voice condemn and completely reject the concept of indiscriminate

Figure 8.3 The camera pans to a large turnout of audiences

killings', there is a change of camera angle and we see a large turnout of audiences following Yusuf on a TV screen. This is followed by a close-up shot in which we see a cameraman recording Yusuf on stage, highlighting the mediation of the event. The subtitling of the sermon in Arabic further suggests a border-crossing popularity for Hamza Yusuf, apart from his English-speaking audiences (also see Schmidt [2005] on Hamza Yusuf as a transnational figure).

If the authority of Hamza Yusuf emanates from his representation, *inter alia*, as a preacher and an intellectual, Baba Ali's authority emanates from representing himself not as being different from his audience but as being similar to them; this shows in his expression of ordinariness and creating conversational interactions with his audience.

The Self-Representation Strategies of Baba Ali

If the performances of Hamza Yusuf are staged in lecture halls, with cameras showing massive co-present audiences, Baba Ali does not have co-present audiences to foreground in his performance. His sermons are mediated on YouTube; he presents his show, while sitting in an armchair against a grey wall. The ordinariness of Baba Ali's performance is represented, however, as a virtue, an aspect he boasts about. In one of his earliest video blogs 'distractions during salat/praying' (broadcast July 2006), Baba Ali starts his video blog by expressing his astonishment that he has got 30,000 views:

Example 3
Thirty thousand views (camera close-up) Thirty thousand views? That's crazy hah. Are all these people watching me doing a simple show like this? That's just me, sitting in a chair, with the camera, and talking about the stuff that's going through my head. But people seem to be

watching (close-up shot) (laughing) some people even laughing. Do you see? I don't know why some people think I am a comedian. For the record, I am not a comedian, I am not an actor. I am just a brother with a video-camera talking about just random stuff.

On the linguistic level, the ordinariness of Baba Ali in Example 3 is discursively constructed through his reference to the staging, i.e. 'that's just me, sitting in a chair, with the camera', and through his repetition of the adverb 'just', i.e. 'that is just me' and 'talking about just random stuff', indicating that he is no different from his audiences. The ordinariness of his representation is also communicated through the use of conversational language, e.g. expressions of personal feeling and conversational interjections (e.g. 'That's crazy hah'), which gives the impression that Baba Ali is speaking his own mind. This is a style that is different from the sermons of Hamza Yusuf, which are staged events where he comes across as having thought about the topics he is discussing.

The conversational nature of Baba Ali's performance is also reflected in the way that he attempts to involve his audiences on YouTube. This is achieved through the use of rhetorical questions (e.g. 'thirty thousand views?') which suggests a shared view of experience (Myers, 2010: 83). In addition, 'Do you see?' is an example of 'enacting conversational interaction' (Myers, 2010: 84–86) in which Baba Ali is engaged in a dialogue with an imagined individual.

Thus, an important strategy through which Baba Ali projects 'informality' and 'ordinariness' is enacting conversational interaction. Recurrent phrases in his video blogs are 'you know what I am talking about, man' (e.g. Culture Versus Islam) and 'Do you know what I am saying?' (e.g. Who hijacked my religion?). These phrases are accompanied by extreme close-up shots where Baba Ali comes closer to the camera, and to the viewer (Figure 8.4).

Figure 8.4 Baba Ali (coming closer to the camera): Do you know what I am saying?

An aspect related to the ordinariness of Baba Ali's self-representation is that the audience is frequently represented as one voice/character in his video blogs. In the video blog 'Distractions during praying', Baba Ali acts the character of an imagined viewer. The excerpt lasts for four minutes in which Baba Ali engages in a self-mocking debate as to whether he can be labelled a comedian. Most importantly, at the end of the video blog, Baba Ali calls on his audiences to take part in the production of his videos. In the following extract, 'Baba Ali 1' represents his persona, whereas 'Baba Ali 2' is that of an imagined viewer.

Example 4

Baba Ali (1): Sometimes you have to be serious.

Baba Ali (2) (frowning like a child): I don't like you when you are serious.

Baba Ali (1): What if I am joking around, you would not take me seriously?

Baba Ali (2): Fine, Okay.

Baba Ali (1): Anyway, this whole video-blog was an experiment; it was never meant a weekly show.

Baba Ali (2): But it is a weekly show, right?

Baba Ali (1): Kind of (.) There is only ten episodes and this is the sixth of the tenth, which means that we have four more left.

Baba Ali (2): What?

Baba Ali (1): Yeah. Four more left. After that I will take a break for a while. I have other stuff to do.

Baba Ali (2): What?

Baba Ali (1): I came to this every week. These things take time to do.

Baba Ali (2): COME ON.

Baba Ali (1): You know what? What about you try it?

Baba Ali (2): What do you mean I try it?

Baba Ali (1): How about you guys put the material together and I will do the show.

Baba Ali (2): Are you serious?

Baba Ali (1): Yeah. You will do the show based on what you think. The audience has many suggestions and comments. Yes, I read them. Sure we can come up with something together. You guys put the content together. I have the camera. I have everything. We can do it. We will broadcast it on YouTube. We will broadcast it on Umma Films and it will go there and you will get comments from everyone.

Baba Ali (2): You serious. Aren't you?

Baba Ali (1): Yeah. I am serious.

Baba Ali (2) (touching his chin with his hands): Okay (.) mhm. Let us do it.

In the above example, the pronoun 'you' is used to address 'the built-in' audience who – idiosyncratically – spoke like a child, signalled by his frown and the tone of his voice. This ties in with Coupland's (2011: 580) observations on performance in popular songs in which the audience accepts that the performer projects a 'persona that audiences know is a character, as opposed to the performer singing in *propria persona as* him/herself'. Towards the end of the dialogue, the reference is broadened to 'you guys', a collective informal reference to the audience as a group, and the inclusive 'we' that sets Baba Ali and the viewers in the same group, e.g. 'sure we can come up with something together'.

An interesting aspect in the above example is Baba Ali's call on his audience to 'put the content together' and take part in the production of his videos. Although Baba Ali's invitation for the audience to take part in the production of his videos appears to be rather rhetorical, since it is unlikely to actually result in any audience participation, it suggests that Baba Ali seems to play with the notion of 'participatory culture' in which 'fans and other consumers are invited to actively participate in the creation and circulation of new content' (Jenkins, 2006: 290).

If Baba Ali differs from Hamza Yusuf in the use of informal register (e.g. 'Ok' and 'you guys'), an aspect that he shares with Yusuf is that – like him – he has multifaceted aspects to his self-representation. He plays the role of a social counsellor who gives advice on personal relationships, a religiously motivated speaker, and on other occasions, he represents himself as an entrepreneur.

An example of Baba Ali giving social advice is his video blog '25,000 Muslim Weddings' (broadcast June 2006) in which he criticises parents who ask for expensive weddings for their daughters/sons. Playfulness appears to be a salient feature in Baba Ali's performances. In the following excerpt, taken from the same sermon, Baba Ali plays two characters: Baba Ali himself (referred to as Baba Ali 1) criticising the fact that some parents stipulate a fancy wedding, and an imagined character (also representing the virtual audience on YouTube) who is being invited to a fancy wedding (Baba Ali 2).

Example 5
Baba Ali (1) (wearing a white t-shirt): For those who never attended one of those fancy weddings, let me break it down for you so that I can tell what you are missing.
Baba Ali (2) (wearing a suit and a tie): Imagine sitting alone at a table in a fancy hall, wondering where everyone is (Baba Ali seems like he is sweating, his hand touching his forehead), quickly learning that 6 pm on the invitation means 7 pm (looking at his watch), getting dressed up in an uncomfortable clothes, sitting in a huge hall and saying 'Wow this must be expensive', getting free refills of soda, eating fancy

food (holding a plate while moving his jaws), staring at the wedding program, saying *salam* to the people you only see at weddings, funerals and *eid*, playing with your napkin, taking pictures with the bride and groom, here is pictures by family and friends crying and saying goodbye (Baba Ali clapping). By the way (2) (looking at the camera with a suit and the tie) why are you saying good bye, they are not dying; they are just getting married.

An element to note in the above video excerpt is the dramatic performance of Baba Ali who performs the actions that are being commented on; for instance, while uttering 'imagine sitting at a table in a fancy hall', Baba Ali rolls his eye upwards to show the character's astonishment of the big fancy hall where the wedding is being held (Figure 8.5).

Following the rhetorical structure of problem/solution, where he gives personal advice, Baba Ali proposes an alternative to expensive wedding ceremonies:

Example 6
But there is an alternative. Try this. Google the web to find inexpensive wedding invitations. Well, you can really save some money. How about having your wedding in the best location? The *masjid* (mosque). The *masjid* (we see the photo of a mosque). You will save a group of money and what honour getting married at Allah *subhanoho wa ta3ala* house. Consider the time between *marrib* to *isha*. That is when people get on their time. Can you imagine the reward of having hundred people to the *masjid* to pray *salat*. That is what I am talking about man. Contact your local halal restaurant and ask him for a deal (Baba Ali holding the phone, talking in a foreign accent) you Know how they say okie dokie, so give me a good deal, hah, I love it I love it mwah. And finally, make *dua* (praying) to Allah *subhano wa talah* and make it pure intention.

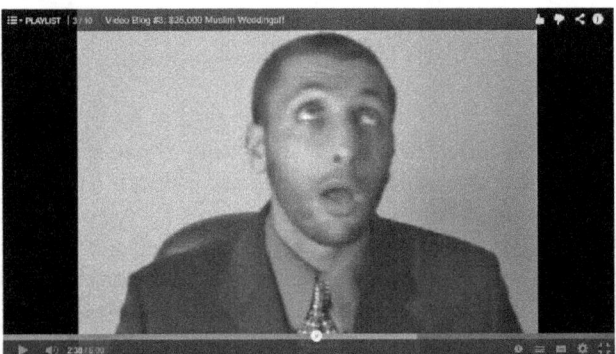

Figure 8.5 Baba Ali acting out his astonishment of the big fancy hall

As in advertisements in which questions and directives are used to engage the audience (see Myers, 2010: 82), Baba Ali, in an example of generic intertextuality, makes use of directives and questions, as if creating an advertisement: 'Try this'; 'Google the web'; 'How about having your wedding in the best location?'

An interesting aspect is that practical solutions are mixed up with religious motivation; the (moral) benefit of having a hundred people pray in the mosque is mentioned as one reason why the mosque is a better place for a wedding to take place. In this way, Baba Ali creates a hybrid fabric of discourse on the social/personal counselling genre and the sermon genre.

There are other sides to this discoursal hybridity (e.g. Chouliaraki & Fairclough, 1999: 83; Fairclough, 1995: 88–90; Wodak *et al.*, 2013). One is that many registers are represented in the video blog; including that of a character calling a local restaurant and asking for a good deal in which specific slang expressions are used (e.g. *'Okie Dokie'* and 'love it'). In addition, the use of spontaneous interpolations (e.g. 'wow') and informal address terms (e.g. 'That's what I am talking about, man') are used. These features – in the above extract – render it not only a hybrid fabric of colloquialism and religious themes but also a space for creativity and playfulness where a polyphony of voices and discursive features (e.g. ethnic accents, colloquial terms and religious terms) are represented. This lends truth to recent research on YouTube that refers to it as a site of 'vernacular creativity' (e.g. see Burgess, 2006, 2007), i.e. 'creative practices that emerge from non-elite, specific everyday contexts' (Light *et al.*, 2012: 433). This in turn indicates the impact of the medium on religious expression online; in other words, the use of YouTube as a platform for religious mediation appears to have given rise to this hybrid genre of colloquialism, religious discourse and comedic sketches in Baba Ali's video blogs.

Playfulness is – in addition – communicated through the use of cinema and vocal techniques. For instance, at the beginning of the video, the phrase 'twenty five thousand dollars' – referring to an estimate of wedding expenses – is mentioned twice: once by Baba Ali, and once by adding a 'robot voice effect' in slow motion, foregrounding the number 'twenty five thousand dollars'. In addition, images are inserted in the video; for instance, as he utters his directive 'Google the web to find inexpensive wedding invitations', there is a cut to an image of wedding invitations on a pink background.

On other occasions, Baba Ali plays the role of the entrepreneur. The entrepreneurial persona of Baba Ali is partly constructed through references to the projects in which he is involved. One of his YouTube videos features an interview with him about a game he created entitled 'Mecca to Medina'. The entrepreneurial personality of Baba Ali is also constructed through creating narratives about characters who would like to have projects with him. For example, in his YouTube video 'my culture', Baba Ali acts the character of a 'cardiologist' who would like to work with him; the aim of the YouTube video is to criticise some 'cultural Muslims who do things that contradict with Islam in the name of their culture', branding it 'Islamic'.

Relevant to the above is a blog post by Baba Ali,[10] where he advertises an event organised by him to help Muslims create networks and partnerships. The blog post starts with a narrative in which Baba Ali recounts his story of meeting young investors who – like him – had 'saved a little money but had nowhere to invest it'. Describing himself in the post as 'the guy with the idea/project' who made partnerships with his friends, Baba Ali moves to the aim of the post which is to sell tickets for an event 'to put together a room full of talented people and ambitious people so they can expand their networks, build connections, and to connect for present or future projects'. Baba Ali ends his post with the claim 'I have learned that it's not always what you know but often who you know that opens doors of opportunity'. The creation of business-networking events by Baba Ali recalls a remark put forth by Miller (2004: 86) about the rise of 'therapeutic' religion, which 'the middle classes in particular need, to placate their contradictions and to assist them in a social world reconfigured in terms of market competition'.

In the same post, Baba Ali refers to his travels, positioning himself as a celebrity. This aspect is further emphasised in his 'stand-up' comedy website,[11] which has a section for Baba Ali's tour dates in the United Kingdom, the United States, Canada and Australia.

The above-examined features, i.e. the ordinariness of his appearance, the characterisation of the audiences and the conversational nature of Baba Ali's linguistic performance are therefore, important strategies that have contributed to Baba Ali's 'style' as an ordinary funny guy unlike the serious tone of Hamza Yusuf that contributes to his representation as an intellectual.

Conclusion

To sum up, I started this chapter by situating this study in the discipline of CDS. I argued that the many aspects that are involved in televangelism (e.g. entertainment elements) call for a critical approach to the investigation of this phenomenon and the reasons that may have contributed to its popularity. Drawing on social theories by Kellner (2003) and Thompson (1995), I proposed – in the second section of the chapter – that televangelism can be seen within the contextual prism of media power and the rise of info-tainment. Furthermore, digital media technologies have offered a platform where many voices can now compete for authority. I proposed that a synergy of DHA (Wodak, 2001) and the theory of Visual Grammar (Kress & Van Leeuwen, 1996) can enable us to deconstruct the performance of televangelists and their self-representation strategies (also see El Naggar, 2015). Due to space limitations, I chose to focus in this chapter on the institutional context, the discourse topics and modes that each televangelist invokes such as dress code, staging and gesture.

The analysis of the sermons of the two televangelists gives us insights into this novel type of religious celebrity who is able to play multifaceted roles. Hamza Yusuf, for example, represents himself, *inter alia*, as a preacher, a media celebrity and an intellectual. Baba Ali represents himself as an entrepreneur, a celebrity and a social counsellor. Moreover, I have shown that Baba Ali and Hamza Yusuf represent two different types of religious celebrities. On the one hand, Hamza Yusuf's authority emanates from representing himself as being different from his audiences, e.g. he is an intellectual. This partly shows in the hybrid discourse topics he invokes on religion and politics and in the modern clothes that he employs in some of his performances, e.g. a suit and tie. Baba Ali's authority, however, emanates from representing himself as being one of his audience. On the visual level, Baba Ali wears informal attire (t-shirt) vis-à-vis the suit usually worn by Hamza Yusuf. On the discursive level, Baba Ali's ordinariness is realised through the use of colloquialisms, conversational interjections and informal address terms. He also uses other means to create a rapport with his audience, for example, through using close-up shots and creating dialogues with an imagined viewer. In this way, Baba Ali's video blogs could be perceived as an example of 'vernacular creativity' in the religious sphere as his sermons feature many clusters and fragments of discourses of slang, colloquialisms and religious expressions.

The analysis further indicates that contemporary religious discourse is open to many hybrid discourses; this shows in Baba Ali's insertion of advertisements into some of his YouTube sermons and/or his drawing on linguistic realisations that are typical of advertisements (e.g. use of rhetorical questions and imperative). It also shows in Hamza Yusuf's convergence towards political rhetoric.

In addition, the analysis of the self-representation strategies of the two televangelists indicates that the two televangelists – invariably – draw on a variety of multiple modes including the use of dress code, graphic elements, camera techniques (e.g. long shots), images and music. Thus, we can conclude that religion has gone down the route of 'info-tainment'. In line with 'politico-tainment' (e.g. Van Zoonen, 2005; Wodak, 2009), it is now possible to coin the term 'religio-tainment' to refer to the manifold manifestations of the blurring between religion and entertainment that are emerging online, for example, televangelism, religious songs, religious hip-hop bands and stand-up comics. The emergence of these genres online seems to offer a promising sub-terrain of research that could be explored further.

Acknowledgement

I am grateful for the insightful comments and remarks of the book editor, Dr Andrey Rosowsky at Sheffield University. I would also like to thank

Distinguished Professor Ruth Wodak and Dr Shuruq Naguib at Lancaster University for their feedback while developing this study. Thanks are also due to the anonymous reviewer for his/her critical comments.

Notes

(1)   Baba Ali's video-channel. See https://www.youtube.com/user/ummahfilms (accessed 30 November 2016).
(2)   Hamza Yusuf's facebook page. See https://www.facebook.com/ShaykhHamza Yusuf/?fref=ts (accessed 30 November 2016).
(3)   For instance, see Gilliat-Ray (2010: 166) on Hamza Yusuf.
(4)   According to the website of Zaytuna college. See http://www.zaytunacollege.org/about/ (accessed 30 November 2016).
(5)   According to Hamza Yusuf's website. See http://sandala.org/about/hamzayusuf/ (accessed 30 November 2015).
(6)   Mandaville (2001:182) refers to 'critical Islam' as a 'notion that has gained particular currency in recent years among diasporic Muslim intellectuals in the West'.
(7)   The article is published in many websites and forums, for instance, see http://www.tikkun.org/article.php/Yusuf-WhyHolocaustdenialunderminesIslam/print (accessed 30 November 2016).
(8)   For example, see his YouTube video 'between ignorance and extremism'. See http://www.youtube.com/watch?v=ryPX3ZIPTjk (accessed 30 November 2016).
(9)   For the transcription of Arabic excerpts, the present study uses the morphophonemic transcription system adapted from Harrell (1957), Hafez (1991) and IPA.
(10) Baba Ali's blog. See http://ummahfilms.blogspot.co.uk/ (accessed 30 November 2015).
(11) Baba Ali's stand-up comedy website. See http://www.muslimcomedian.com/ (accessed 30 November 2016).

References

Blackledge, A. (2005) *Discourse and Power in a Multilingual World.* Amsterdam: John Benjamins Publishing Company.
Bruce, S. (1990) *Pray TV: Televangelism in America.* London: Routledge
Burgess, J. (2006) Hearing ordinary voices: Cultural studies, vernacular creativity and digital storytelling continuum. *Journal of Media and Culture Studies* 20 (2), 201–214.
Burgess, J. (2007) Vernacular creativity and new media. PhD thesis, University of Queensland.
Butler, J. (1990) *Gender Trouble: Feminism and the Subversion of Identity.* London/New York: Routledge.
Campbell, H.A. (2010) *When Religion Meets New Media.* London: Routledge.
Campbell, H.A. (2013) *Digital Religion: Understanding Religious Practice in New Media.* Abington: Routledge.
Chouliaraki, L. and Fairclough, F. (1999) *Discourse in Late Modernity: Rethinking Critical Discourse Analysis.* Edinburgh: Edinburgh University Press.
Couldry, N. (2000) *The Place of Media Power: Pilgrims and Witnesses of the Media Age.* London: Routledge.
Couldry, N. (2003a). *Media Rituals: A Critical Approach.* London; New York: Routledge.
Couldry, N. (2003b) Television and the Myth of the Mediated Centre: Time for a Paradigm Shift in Television Studies? Paper presented at the Media in Transition 3 Conference, MIT, Boston, 2–4 May, 2003.

Coupland, N. (2011) Voice, place and genre in popular song performance. *Journal of Sociolinguistics* 15 (5), 573–602.

Echchaibi, N. (2011) From audio tapes to video blogs: The delocalization of authority in Islam. *Nations and Nationalism* 17 (1), 25–44.

El Naggar, S. (2012) Intertextuality and interdiscursivity in the discourse of Muslim televangelists: The case study of Hamza Yusuf. *Critical Approaches to Discourse Analysis Across Disciplines* 6 (1), 76–95.

El Naggar, S. (2014) The impact of digitization on the religious sphere: Televangelism as an example. *Indonesian Journal of Islam and Muslim Societies* 4 (2) 189–211.

El Naggar, S. (2015) Multimodality in perspective: Creating a synergy of the discourse historical approach and the framework of visual grammar. In J. Wildfeuer (ed.) *Building Bridges for Multimodal Research: International Perspectives on Theories and Practices of Multimodal Analysis* (pp. 149–166). Bern/New York: Peter Lang.

El-Nawawy, M. and Khamis, S. (2009) *Islam Dot Com: Contemporary Islamic Discourses in Cyberspace*. New York: Palgrave Macmillan.

Esposito, J. and Kalin, I. (eds) (2009) *The 500 Most Influential Muslims in the World*. Amman: The Royal Islamic Strategic Studies Centre.

Fairclough, N. (1995) *Media Discourse*. London/New York: E. Arnold.

Fairclough, N. (2003) *Analysing Discourse: Textual Analysis for Social Research*. New York: Routledge.

Fairclough, N.L. and Wodak, R. (1997) Critical discourse analysis. In T.A. van Dijk (ed.) *Discourse Studies: A Multidisciplinary Introduction, Vol. 2. Discourse as Social Intercation* (pp. 258–84). London: Sage.

Giddens, A. (1990) *The Consequences of Modernity*. Cambridge: Polity Press in association with Basil Blackwell, Oxford, UK.

Gilliat-Ray, S. (2010) *Muslims in Britain: An Introduction*. Cambridge: Cambridge University Press.

Goodwin, C. (2003) Pointing as situated practice. In K. Sotaro (ed.) *Pointing: Where Language, Culture and Cognition Meet* (pp. 217–241). Mahwah, NJ: Lawrence Erlbaum.

Goffman, E. (1959) *The Presentation of Self in Everyday Life*. Harmondsworth: Penguin.

Hafez, O.M. (1991) Turn-taking in Egyptian Arabic: Spontaneous speech vs. drama dialogue. *Journal of Pragmatics* 15, 59–81.

Harrell, R.S. (1957) *The Phonology of Colloquial Egyptian Arabic*. New York: American Council of Learned Societies.

Hewitt, N. (2003) *The Cambridge Companion to Modern French Culture*. Cambridge: Cambridge University Press.

Hinnells, J. (2010) The study of diaspora religion. In J.R. Hinnells (ed.) *The Penguin Handbook of the World's Living Religions* (pp. 684–691). London: Penguin.

Hoesterey, J. (2008) Marketing morality: The rise, fall, and re-branding of Aa Gym. In G. Fealy and S. White (eds) *Expressing Islam: Religious Life and Politics in Indonesia* (pp. 95–112). Singapore: Institute of Southeast Asian Studies.

Howell, J.D. (2008) Modulations of active piety: Professors and televangelists as promoters of Indonesian Sufisme. In G. Fealy and S. White (eds) *Expressing Islam: Religious Life and Politics in Indonesia* (pp. 40–62). Singapore: Institute of Southeast Asian Studies.

Jenkins, H. (2006) *Convergence Culture: Where Old and New Media Collide*. New York: New York University Press.

Krauss, R.M., Chen, Y. and Gottesman, R.F. (2000) Lexical gesture and lexical access: A process model. In D. McNeill (ed.) *Language and Gesture* (pp. 261–283). Cambridge/New York: Cambridge University Press.

Kress, G. and van Leeuwen, T. (1996) *Reading Images: The Grammar of Visual Design*. London: Routledge.

Kellner, D. (2003) *Media Spectacle.* London/New York: Routledge.

KhosraviNik, M. (2010) Self and other representation in discourse: A critical discourse analysis of the conflict over Iran's nuclear programme in the British and Iranian newspapers. PHD thesis, Lancaster University.

Light, B., Griffiths, M. and Lincoln, S. (2012) Connect and create: Young people, YouTube and graffiti communities. *Continuum* 26 (3), 343–355.

Lotfy, W. (2009) *The Phenomenon of Televangelists.* Cairo: Dar El Ein.

Mandaville, P. (2001) *Transnational Muslim Politics: Re-imagining the Umma.* New York: Routledge.

Mandaville, P. (2007) *Global Political Islam.* London: Routledge.

McNeill, D. (1992). *Hand and Mind: What Gestures Reveal about Thought.* London: University of Chicago Press.

Messinger, S.E., Sampson, H. and Towne, R.D. (1975) Life as theater: Some notes on the dramaturgic approach to social reality. In D. Brissett and C. Edgley (eds) *Life as Theatre: A Dramaturgical Sourcebook* (pp. 32–42). Chicago, IL: Aldine Publishing Company.

Miller, V.J. (2004) *Consuming Religion: Christian Faith and Practice in a Consumer Culture.* New York: Continuum.

Myers, G. (2010) Stance-taking and public discussion in blogs. *Critical Discourse Studies* 7 (4), 263–275.

Reisigl, M. and Wodak, R. (2009) The discourse historical approach. In R. Wodak and M. Meyer (eds) *Methods of Critical Discourse Analysis* (pp. 87–121). London: Sage.

Richardson, J.E. and Wodak, R. (2009) Recontextualizing fascist ideologies of the past: Right-wing discourses on employment and nativism in Austria and the United Kingdom. *Critical Discourse Studies* 6 (4), 251–267.

Riegert, K. (2007) The ideology of *The West Wing:* The television show that wants to be real. In K. Riegert (ed.) *Politicotainment: Television's Take on the Real* (pp. 213–236). Bern: Peter Lang.

Robinson, F. (1993) Islam and the impact of print. *Modern Asian Studies* 27 (1), 229–251.

Schmidt, G. (2005) The transnational Umma – myth or reality? Examples from the western diasporas. *The Muslim World* 95, 575–586.

Schmidt, R. and Kess, J.F. (1986) *Television Advertising and Televangelism: Discourse Analysis of Persuasive Language.* Amsterdam/Philadelphia, PA: John Benjamins.

Stöckl, H. (2009) The language-image-text-theoretical and theoretical inroads into semiotic complexity. *AAA – Arbeiten aus Anglistik und Amerikanistik* 34 (2), 203–226.

Swartz, M. (1999) Arabic rhetoric and the art of the homily. In R.G. Hovannisian and G. Sabagh (eds) *Religion and Culture in Medieval Islam* (pp. 36–59). Cambridge: Cambridge University Press.

Tartoussieh, K. (2013) Muslim digital diasporas and the gay pornographic cyber imaginary. In E. Alsultany and E. Shohat (eds) *Between the Middle East and the Americas: The Cultural Politics of Diaspora* (pp. 214–230). Ann Arbor, MI: University of Michigan Press.

Thompson, J.B. (1995) *The Media and Modernity.* Oxford: Polity Press.

Van Leeuwen, T. (2011) *The Language of Colour: An Introduction.* London: Routledge.

Van Zoonen, L. (2005) *Entertaining the Citizen: When Politics and Popular Culture Converge.* New York: Rowman and Littlefield.

Wise, L. (2003) Words from the heart: New forms of Islamic preaching in Egypt. Master's thesis, St Anthony's College, University of Oxford.

Wodak, R. (2001) The discourse historical approach. In R. Wodak and M. Meyer (eds) *Methods of Critical Discourse Analysis* (pp. 63–94). London: Sage.

Wodak, R. (2010) The glocalization of politics in television: Fiction or reality? *European Journal of Cultural Studies* 13 (1), 43–62.

Wodak, R. (2011) *The Discourse of Politics in Action: Politics as Usual.* Basingstoke: Palgrave Macmillan.

Wodak, R. (2015) *The Politics of Fear: What Right-Wing Populist Discourses Mean.* London: Sage.

Wodak, R. and Meyer, M. (2009) Critical discourse analysis: History, agenda, theory and methodology. In R. Wodak and M. Meyer (eds) *Methods of Critical Discourse Analysis* (pp. 1–34). London: Sage.

Wodak, R., Khosravinik, M. and Mral, B. (eds) (2013) *Right-Wing Populism in Europe: Politics and Discourse.* London: Bloomsbury.

Part 4

Faith, Language and Online Ritual

9 Online *Satsang* and Online *Puja*: Faith and Language in the Era of Globalisation

Rajeshwari V. Pandharipande

This chapter explores online Hindu religious practices in the US diaspora. The two Hindu rituals in focus are *Satsang* (literally, connecting through religious discourses with the Divine/Gurus who have the knowledge and experience of the Divine) and *Puja*, the ritual worship of Hindu deities. These online ritual practices are on the rise in the US. The two rituals mentioned are central to the religious experience of Hinduism. Online contact with the Divine presents a dilemma. The question that faces the theological tradition is whether virtual contact mediated by technology equates with face-to-face contact (for producing the experience). Related to this situation, one may ask the question in the Zen context, 'Can Zen master conduct Koan-session online? Does mind to mind transmission take place in the mediated contact?' This chapter will compare online (virtual) and actual (physical) rituals in the context of the nature of contact with the Divine in the two methods. The data will be drawn from the following sources: devotees who conduct online rituals; Gurus who carry out *Satsang* online; the history of Hinduism which has undergone many changes (for example, the change of mode in religious communication from oral to written). This chapter will also include discussion on the questions of authority (Pandharipande, 2010) in the context of new methods of communication.

Introduction

The two major hallmarks of the 21st century are technology and globalisation. The use of digital media in communication is indisputably accepted in all walks of life such as sports, medicine, music, education, aviation, etc. Its legitimacy and/or authenticity are not questioned in these contexts. In fact, digitisation is viewed as the most powerful method to preserve and spread knowledge and information in both secular and religious realms. Sacred religious texts are now being digitised to facilitate their preservation and access from any part of the world. However, the legitimacy of online/digital religious rituals has been questioned and is currently being debated both in academic discussions and in the domain of religious practice. In the corporate business context, a digital PowerPoint presentation wields much more power and is considered more effective than a presentation without it. However, online rituals such as online worship of deities or online conversation/discussion with Gurus are not readily accepted as legitimate or authentic religious practices. Moreover, they are viewed (by many priests and practitioners of rituals) as lacking the efficacy of their offline counterparts. Nonetheless, with the fast growing use of digital media, increasingly more Hindus in the US diaspora are beginning to use digital media for carrying out or performing their rituals. This chapter explores two online Hindu religious practices in the US diaspora. They are, *satsang* (literally, connecting through religious discourses with the Divine/Gurus who have the knowledge and experience of the Divine) and *puja*, the ritual worship of Hindu deities. These online ritual practices are on the rise in the US. These two rituals are central to the religious experience of Hinduism. Traditionally, it is assumed that *darshan* (direct vision) of the Divine (which is part of the above two rituals) is one of the most powerful methods for the religious experience. It is believed that a Guru transfers Knowledge to the aspirant during *satsang* and the deities are 'carved' (permanently established) in the heart of the devotee in the *puja* ritual (Eck, 1985).

The ensuing discussion aims at accomplishing the following five major goals. First, I will analyse the form and function of *satsang* and *puja* rituals in the Hindu tradition and compare those rituals with their online counterparts to identify the difference between the two; second, I will examine the role of language in both its online and offline form and function; third, I will discuss the notions of authenticity and authority in the Hindu tradition and the perspectives of the devotees and scholars on the authenticity and legitimacy of online rituals; fourth, I will argue that within the Hindu framework of rituals, the change/difference in online and offline rituals should not be held responsible for marking them as unauthentic, or lacking efficacy; finally, I will propose that when the method of performing/practicing rituals changes (from offline to online in this case), the physical form/structure of the ritual changes which in turn changes the relationship between signifiers (the objects in the ritual) and the signified (the religious meaning). This relationship is generally not readily accepted within the religious community until it is conventionalised. However, once the new relationship is well conventionalised through time and by the 'authority' (saints and Gurus in this case), the online rituals are acceptable as legitimate.

I will point out that the change and variability in the medium and method of carrying out rituals and the acceptance of their legitimacy over time has been part of the history of Hinduism.

The analysis of online/offline rituals presented here is based on many sources: the study of online rituals; direct observation of their performance in temples, homes and public places in the US and India; and many discussions with the devotees/believers and saints. Some of the discussions were part of informal interviews. Hinduism does not have institutions similar to the church in the Christian tradition, which is seen as an authority to validate changes in the tradition. Since saints/Gurus are viewed as authority in the Hindu tradition, I valued discussions with saints as particularly important. Moreover, discussions/conversations and informal interviews with the devotees or practitioners of the rituals proved valuable in order to understand the response and reception of the Hindus to this recent phenomenon of online rituals. I decided to opt for long informal discussions as opposed to formal interviews at this point in time, as I found people more willing to discuss many topics such as the form and function of rituals as opposed to responding to formal questions. This interaction gave them a chance to share their views on the past changes in rituals in the tradition.

Perspectives on religious rituals

Before I discuss online rituals, it is necessary to provide a brief review of various perspectives on rituals presented within the frameworks of anthropology and religious studies and point out my approach to the analysis of online rituals. Definitions of religious rituals presented in Durkheim (1995), Geertz (1973), Leach (1968), Malinowski (1954) and Turner (1969), among others, show variation. The two dominant approaches to ritual are observed in Malinowski (1954) and Durkheim (1912/1995). For Malinowski, ritual is the method or a way of dealing with situations which create fear and anxiety and which people perceive as beyond their control. Thus, rituals have a psychological motivation behind them. For example, according to Malinowski, the islander fishermen perform rituals to deal with the fear caused by the situation of deep-sea fishing. When they fish in a safe environment, they do not perform rituals. Durkheim (1995) treated ritual as an activity to maintain the sense of collective identity of the community and a sense of belonging. According to Durkheim, rituals are important means of maintaining society. For example, by regularly performing religious rites at a certain time and space, people develop a strong sense of membership in a larger community. While Malinowski and Durkheim define rituals by their function, i.e. as responses to the psychological and social needs, respectively, Turner (1969) and Leach (1968) do not subscribe to defining rituals based on one single function. They argue that ritual has many forms and functions. Turner, following Gennep's work, treats ritual as social interaction carried out through conventionalised symbols. Geertz (1973) focuses on the performative and creative function of ritual and defines ritual as a system of symbols that provides both a *'model of'* reality (key for the interpretation of the world) and a *'model for'* the reality (ideal state of reality/world). Helland (2012: 27) consolidates various approaches under two perspectives on ritual – *what it is* and *what it does*. The former perspective describes the ritual performance (what it is) while the latter describes its function (what it does). However, as Helland points out, these two are not always separated as in McGuire's (1997:16) definition, 'ritual consists of symbolic actions that represent religious meaning'. Helland's own definition of ritual as 'purposeful engagement with the sacred (whatever the sacred may be for those involved)' is broad and accommodates various perspectives on ritual. According to Kreinath *et al.* (2004, in Helland, 2012: 28–29), 'religious ritual is an aggregate of performance, media, script, and representation of belief'. In all of the definitions above, it is assumed that religious ritual is an intentional activity which is authenticated by tradition. Kreinath's definition is comprehensive; it integrates different characteristics of ritual and is useful for understanding the difference between online and offline rituals. Helland elaborates on the terms 'script', 'performance', 'media' and 'representation of beliefs'.

> ...the script is the set of rules that are laid out to be followed by the ritual, the words, action, gestures, and symbols that are to be used. The performance is the carrying out of the ritual, the attempt to follow the script. The media are the mechanisms for communicating and receiving the ritual. Last but not the least, are the representations of belief that are embodied through the ritual performance itself: myths, sacred narratives, sacred stories, the belief of the supernatural or whatever the participants view as sacred. (Helland, 2013: 27–28)

In the ensuing discussion on the two Hindu rituals, I rely on the following salient features of rituals mentioned in the above definitions, i.e. a ritual has a 'script' (a religious ritual has a set of rules to follow), there is a mechanism through which it is performed, it symbolically represents religious beliefs (Kreinath *et al.*, 2004) and it is a 'purposeful engagement with the sacred' (Helland, 2013). Additionally, I assume, following Geertz (1973), that religious ritual is a system of symbols that conveys religious meaning.

Researchers also observe that rituals change according to their socio-religious context. Although the above aspects of ritual can be seen as building blocks of a religious ritual, it is important to keep in mind these various aspects of rituals in order to fully understand the two online rituals under focus – *satsang* and *puja*. My goal is to identify the form and function of online rituals and point out that the questions of efficacy, authenticity and authority cannot be discussed independently of the beliefs in the Hindu tradition and their sociocultural context.

Language in online rituals

This short section focuses on the role of language in online rituals. This discussion is useful to understand the differences between online and offline rituals. Although the 'script' of religious ritual mentioned above includes words/language (Kreinath, 2004), scholars have not adequately focused on the role of language in online rituals. Language plays a vital role in online rituals. Two major functions of language are observed in online rituals – the performative function and the communicative function. The first performative function of language is seen when the language 'creates' the sacred space of online rituals. For example, the online rituals are performed in a space (internet/cyberspace) which is new and not traditionally sanctioned. It is the language used in the rituals (*puja*, the worship ritual, for example) which creates the sacred space through communication/instructions on the computer screen for performing rituals such as, 'Now you enter the temple, ring the bell, pick up flowers, offer those to the deity in the temple, light up the incense stick, etc.'. The virtual space symbolically represents the physical space (temple). It is important to note that this virtual sacred space, unlike the physical space, is created anew each time the ritual is performed.

The second important performative function of the language used in rituals (recitation of sacred texts, prayers, etc.) is viewed as a powerful instrument in establishing contact with the Divine. As mentioned above, religious rituals depict the engagement of the devotees/performers of the rituals with the sacred. The devotees believe that language accompanies the actions in the performance of rituals to connect with the Divine/sacred. The third significantly important function of the language is that it contributes to the process of signification of the religious meaning of rituals. As Geertz (1973) points out, religious ritual is a system of symbols through which the 'religious meaning' is conveyed. For example, in the Hindu worship ritual, objects such as language and the material (water, jars, incense, flowers, music, idols, photographs of the deities, etc.) are part of the system of symbols which express the meaning of worship (further discussion in the sections on *satsang* and *puja*). Individually and collectively, they express religious meaning (which in this context is the process of connecting with the sacred). If we accept that a symbol represents something (concrete or abstract beyond itself), then we also assume after Saussure (see Holdcroft, 1991) that a symbol is a sign or a signifier which signifies or indexes meaning beyond itself. In the tradition of semiotics developed by Saussure (referred to as *semiology*), the sign relation consists of a form of the sign (the signifier) and its meaning (the signified). Saussure saw this relation as being essentially arbitrary, motivated only by social convention. For example, in the worship ritual, a flower is a signifier which signifies an offering to the sacred; the idols signify the sacred/Divine, etc. Language is also believed to be a signifier. It signifies/symbolises connections with the sacred. Further discussion on the process of signification in online rituals will be taken up later in the discussion.

Additionally, language used in online rituals has a communicative function in online rituals. That is, it is used to interact/communicate with the religious community. The language or the linguistic code used in online rituals differs from traditional language to make communication/ instructions comprehensible to 'internet religious communities'. This necessitates the choice of an appropriate linguistic code (structure), as well as a method of communication/delivery (the use of appropriate symbols, choice of religious texts and the appropriate mix of languages suitable for the local/global Hindu communities). In this sense, the language in online rituals is flexible. In contrast to this, language in offline rituals is relatively 'fixed' in its linguistic form since the religious community participating in the rituals is expected to be relatively homogeneous and familiar with the rituals and the language(s) used in them. I would like to argue that the language used in online rituals is variable because on the one hand it has to take into account the goals of the virtual community, the level of its familiarity with the rituals and the constraints of the language, but, on the other hand, the language has to abide by the constraints of the internet (for example, certain actions such as walking around a temple which is done by devotees in offline *puja* cannot be done in online *puja*). Thus, the instructions for the performer of the ritual cannot include these instructions to perform actions such as this.

In the following discussion, I will first describe two rituals, *satsang* and *puja*, and compare these with their respective offline counterparts and point out significant differences between the two. Additionally, I will compare online *satsang* with online *puja*. I would like to point out that, in general, devotees (who may or may not perform online *puja* and those who may or may not participate in a *satsang*) prefer face-to-face interaction with the person, generally the Guru (in *satsang*) and the actual/physical (not virtual) 'deities' in the *puja*. In this context, I interviewed a saint at Rishikesh, one of the most important sacred places of Hindu pilgrimage in India. When asked about his preference for online or offline *puja*, the spiritual master (saint) said to me, 'A photograph is not the real person. Similarly, virtual Guru is not the real person, virtual deities are not real deities. Of course, offline *satsang* and offline *puja* are far superior compared to their online counterparts'. However, both the devotees and the spiritual master stated that of the two, online *satsang* is better and more effective than online *puja*. In this context, I will discuss the perspectives of scholars on the issues of the efficacy and authenticity of online ritual along with language and point out that the questions and concerns (empirical and theoretical) raised by scholars can be answered in a straightforward fashion if we take the position that it is lack of establishment/conventionalisation of the signifier–signified relationship that explains why technology-mediated religious rituals are not readily accepted by the practitioners of the religion at this point in time. The discussion will include evidence from the history of Hinduism where the reconfiguration of the signifier and signified relationship has taken place.

The major argument for the perceived relatively lower efficacy/validity of online rituals compared to their offline counterparts is because online rituals have not yet established the signifier–signified relationship between the virtual and physical images. I argue that the relationship between the signifier and signified is established or conventionalised within a culture of a religious community through tradition. Language is an integral part of the whole equation of the signifier and signified. That is, language along with ritual objects, process and the physical context are culturally conventionalised signifiers for the intended signified meaning/goal of the ritual (*puja*, for example). I will further show that the historical change in the language of ritual is accompanied by a cultural change that has reconfigured the relationship between signifier and signified. More discussion will follow on the authenticity of the change in language form and its use. The authenticity/legitimacy of the change in the equation of signifier and signified will also be discussed.

Satsang in Hindu tradition

As mentioned earlier, *satsang* and *puja* are central to the Hindu tradition. These two terms convey diverse meanings in Hinduism. These meanings emerge as a result of an emphasis on different dimensions of the concept of *satsang* and its use in diverse contexts. *Satsanga* is a Sanskrit word (*satsang* in many modern Indian languages where the word-final vowel '/a/' has been lost over a period of time) which etymologically means *sat* 'truth' and *sanga* 'company' – 'In the company of the highest Truth'. Additionally, *sat* can be interpreted as 'a spiritual person or saint who has realised/experienced the Truth' and *sanga* 'company'. In this case, *satsang* means 'in the company of a saint/spiritual person' – in the company of the Guru, a spiritual master, since traditionally the Guru is believed to be the 'realised soul'. *Satsang* is also viewed as an event/gathering where devotees/aspirants meet with the spiritual teacher/saint from whom to seek guidance for their spiritual enhancement. In this context, *satsang* refers to discussions; conversations between the spiritual teacher/saint and the seekers which involve the teacher's instructions; the reading of scriptures; meditation and question-answer sessions between the saint and devotees/aspirants. Thus, *satsang* is treated as a ritual to acquire knowledge about the spiritual goal of life as well as the path/method to reach that goal. As Frisk (2002: 64–85) succinctly points out, '*Satsang* is sitting together with an enlightened person who usually gives a short speech and then answers questions'. While religious activities such as the reading or recitation of scriptures, meditation and worship rituals, individually and collectively, can be seen as *satsang* (since they are all oriented toward reaching the Divine), the Hindu tradition has always emphasised the presence of an enlightened person, saint or Guru as a necessary feature of the phenomenon of *satsang*. It is believed that through direct contact in *satsang*, the Guru transmits the knowledge/experience of enlightenment to the disciples.

The tradition does not claim that there is homogeneity across the methods of communication used by different Gurus in the spiritual meetings neither do all Gurus provide the same instruction. There are literally thousands of records of *satsang* of various types with various saints. While Ramana (one of the most celebrated saints of 20th-century India) used to provide *satsang* in silence (without uttering a word), Sri Sri Ravi Shankar conducts it through a dialogue with disciples in gatherings/meetings/discussions. Additionally, different saints use/teach different scriptures and methods of developing spiritual consciousness. It is important to note here that different saints or Gurus use different languages (as deemed necessary according to the topic of discussion, audience's familiarity with the language of the sacred texts, etc.) The Guru answers questions related to spirituality and religion such as the nature of the Divine, how the Divine relates to the physical world, etc. *Satsang* is considered important for understanding religion, its goal and its relevance to secular life. *Satsang* is also important for clarifying misconceptions and doubts through listening to and interacting with a Guru (the 'realised soul') who is believed to have an understanding of the faith and who has experienced the ultimate reality/Divine/God. The *satsang* is believed to lead the disciples on the spiritual path.

Although *satsang* encompasses a large space of meanings, conventionally, *satsang* generally refers to a gathering of disciples where a saint gives discourses on diverse topics related to religion. Although the primary focus of *satsang* is spiritual/religious knowledge and its necessity, the discourse when interactive, generally includes disciples' questions related to secular/ worldly life, the difficulties/problems therein and the solutions to them based on spiritual knowledge. The following example is typical in a *satsang* with Sri Sri Ravi Shankar, the founder of the Hindu religious mission named *The Art of Living*. The report on *satsang* includes the following discussion on the relationship between knowledge and happiness. I have separated the Guru's speech from the enquirer's and added the translation of the word *seva* (service) for clarification.

Guru: If someone is unhappy, that means they have not received or taken the knowledge; they have not digested the knowledge, and so they are unhappy.

Enquirer: You can say, Gurudev, there are people who have no water, no food and are suffering. That is why they are unhappy. How can they be happy?

Guru: This is a different thing; a different issue. Yes, suffering is different. People suffer because of natural calamities, or due to drought, or flood. In these circumstances, you should do seva (service). Even here, seva (service) with knowledge is much better. ('Niranjani', n.d.)

Traditionally, *satsang* is viewed as an interactive conversation between the saint/Guru and the disciples as in the above dialogue with the Guru who is physically present. The traditional spaces for *satsang* have been temples or public spaces such as lecture halls, open-air theatres or the homes of devotees wherever adequate space is available. The 'live connection in person' between the Guru and the disciples was/is central to *satsang*. In this age of globalisation and technology, *satsang* has acquired a new incarnation. Many methods are adopted by the disciples for *satsang*: (1) unlike offline *satsang*, online *satsang* does not require the physical presence of the saint or the Guru; rather, the disciples or believers collectively or individually watch prerecorded discourses of the Guru. The space is variable such as temples, homes, public spaces, etc. For example, the *Swadhyay*(a), a group of disciples in India as well as in the US, watch prerecorded discourses of Athavale Guruji at home or in a public space. In this case, the disciples listen to the Guru's talks but do not directly interact with the Guru, rather, they interact with each other. In the US, the International Society of Krishna Consciousness (ISKON) also follow the same process of watching prerecorded interviews and discourses by Prabhupada (the founder of the ISKCON). (2) Online interactive *satsang* with the Guru: in this *satsang*, the disciples or other people can interact with the spiritual leader/Guru online (in a virtual space like internet). The conversation can be a discourse by the spiritual master followed by (or during the discourse) questions from listeners and the Guru's respective answers. For example, in the interactive meeting with Bodhisattva Swami Premodaya ('Free Online Satsang', n.d.) detailed instructions are given about how to conduct the interaction with the Guru. The instructions are typical for such interactive *satsang* (Figure 9.1).

In Figure 9.1, it is clear that the audience can interact with the Guru/saint directly through the computer and can see him as well. In this context, the disciples can ask questions and receive answers from the Guru similar to offline *satsang* where the Guru and disciples enter into a dialogue.

LOG-ON INSTRUCTIONS:

The **ONLINE SATSANG** is a live, real-time, interactive online meeting: Your active participation is appreciated, and we invite your questions, reports and comments.

You can join by computer, telephone or mobile device. If by computer or mobile device,

1. Go to:

http://fuze.me/25370867

2. Enter your name.
3. Be sure the "Launch Fuze meeting in the browser" box is unchecked. (on mobile devices a meeting number may be requested, if so, enter[masked])

4. Click "Join as attendee".
5. Select your audio conference preference (VoIP is recommended, if you're using a computer).

6. Click the "Join video conference" button (Note: your camera may be "paused" — be sure to unpause it). To view the video full screen, click on the "expand" button (located near the video screen).

7. You will be muted initially; to speak with Swami Premodaya, raise your flag by clicking on the flag icon, or write to the meeting moderator, by typing in the "Meeting chat" box.

If joining on your telephone,

1. Dial (201)[masked]

2. Enter meeting number: [masked] then press "#".

3. Please 'mute' your phone, and when you wish to speak, unmute yourself (if you don't mute your phone, the moderator will have to mute your line, and won't know when you wish to speak).

4. When speaking, please don't use the speakerphone function.

On the technical side, we need your cooperation in the following ways, for a smoother meeting:

1. Wear headphones (helps prevent audible echo and feedback).

2. Connect via an Ethernet line, as Wifi can be inconsistent and streaming video is bandwidth heavy.

3. Turn off all other programs and applications, so only Fuze is running on your computer.

4. If you are having difficulties, often powering down your computer and re-starting after 2 minutes helps.

Thank you for participating in ONLINE SATSANG.

Figure 9.1 Instructions for participating in the Satsang

Although online *satsang* is on the rise, offline *satsang* is also gaining popularity. Spiritual leaders (Hindu, Sikh, Jain and Buddhist) both in India and abroad travel and conduct offline *satsang* with small and large groups. A majority of saints/spiritual leaders do both online as well as offline *satsang*.

Differences in the language of online and offline *satsang*

Online *satsang*, or connection with the Guru, differs from its live or offline counterpart in many ways. The most important difference is that online *satsang* is not restricted to a particular group of people, time or space. For example, a person can connect with Guru Mooji, a Jamaican (Hindu) Guru, from anywhere in the world. Amritanandamayi, a woman saint of India, can connect with anyone across the world as well. The discourses of these saints can be accessed from anywhere in the world anytime and by anyone. In contrast to this, in the live *satsang*, the community of disciples may place restrictions on the audience, i.e. who can participate in it. Some temple-based *satsangs* do not allow non-Hindus in the *satsang*. In this sense, online *satsang* can be treated as more liberal than its offline/live *satsang*.

There is permanency about online *satsang*. It might be recorded in real time (day, month, year) but once recorded, it can be used by different groups of people at will in their homes, temples or public spaces. In contrast to this, *puja*, as we will see later, has to be performed each time the devotee aspires to connect with the Divine. Its efficacy lies in its actual performance, not in watching its recorded version.

Another major difference is in the language used in the online *satsang* and offline *satsang*. Online *satsang* may or may not have a well-defined homogeneous community with a shared faith since anyone can connect with the Guru and interact. In this sense, online *satsang* is more democratic than its offline counterpart. In contrast to this, in offline live *satsang*, the audience is generally a community with shared beliefs. This difference impacts language use. In online *satsang*, the language used is expected to be understood by a larger, non-homogeneous/diverse community using many languages in different places around the world. Therefore, online *satsang*, though more democratic in its audience selection, is more restricted in language use. In the US, in online *satsang* of Hinduism, Buddhism, Sikhism and Jainism, English is used almost exclusively. In India as well, in the *satsang* of Sri Sri Ravishankar, Dayananda Saraswati and many others, the language is English. In contrast to this, in live offline *satsang*, the Guru uses the language of the community, i.e. Hindi in the north of India, Marathi in Maharashtra, Gujarati in Gujarat, Tamil in Tamilnadu, etc.

It is important to note here that as the size of the communities that use Indian regional languages increases in the diasporas both within India (where communities migrate to states where the majority language is not their own, such as Gujarati-speaking communities moving to Punjab) and outside India (UK, US, Canada, Australia, etc.), the language of online *satsang* changes. That is, two types of languages are adopted by the *satsang* communities (including the Gurus presiding over *satsang*): the English language to include larger (the global Hindu and non-Hindu audience) multilingual communities across the globe and the regional Indian languages (for the communities who speak those particular languages). For example, in the UK, the Dadabhawan TV channel conducts *satsang* in the regional language, Gujarati, and it is translated into Marathi as well. The Dish Channel in the US broadcasts *satsang* in the regional languages of India which are broadcast all over the world. This ability of technology to provide choice of language according to the need of the community/audience has almost eliminated the difference between online *satsang* and its offline counterpart. The question arises then, why is live/offline *satsang* with the presence of a live Guru preferred by many disciples? Is offline *satsang* more authentic? Is the goal of *satsang* better accomplished in the offline *satsang* with the presence of a live Guru? Does face-to-face conversation mirror a technology-mediated one? In the following discussion, I will describe online *puja* and point out how it differs from offline *puja* and then take up these questions in the following section.

Puja in Hindu tradition and online *puja*

There is a large body of literature on Hindu rituals in general and *puja* in particular. However, online *puja* is a new phenomenon similar to online *satsang*. *Puja* is traditionally viewed as a worship ritual. The Sanskrit word *puja* (*puujaa*) is derived from the verb root *puuj* 'to worship'. *Puja*, similar to other religious rituals, can be defined in terms of 'what it is' and 'what it does' (Helland, 2012: 27). From the perspective of the former – 'what it is' – a religious ritual such as *puja* is a conventionalised activity performed repetitively on many occasions by its devotees. It is a performative act through which the devotee connects with the Sacred, whatever God/ Divine (abstract or concrete) is believed to be in the tradition such as a Higher power (human or other incarnations of the Divine, etc.). This is believed to be the function of the *puja* ritual. The 'script' of *puja* includes a range of actions from the simple action of folding hands and praying, to more elaborate actions like offering flowers, fruit, incense and food to the Divine, reciting scriptures, singing *bhajans* (religious prayers put to music), meditation, etc. The worship ritual can (but not necessarily) involve one or many people including a priest who may preside over the ritual on behalf of the devotee(s). One of the main aspects of the ritual is the *murti* or *vigraha* (literally meaning the concrete expression of the Divine), an idol/statue of a Hindu deity to whom the above-mentioned offerings are presented. This ritual is performed at home, in a temple or other spaces depending on the nature and type of worship. Certain *pujas* may be performed anytime and anywhere while others must be performed at a designated time/day/month/ year, etc. After it is offered to the deity, the food is distributed among the people present at the ritual as a blessing (*prasada*) of the Divine. There are many types of *pujas*, the performance of which varies according to the occasion and purpose of the *puja*. For example, *puja* for the god Ganesha involves offering a particular sweet, *modaka*; the worship of Hanuman includes a reading of the text, *Hanumanchalisa*, etc.

The following description elaborates the method/process of accomplishing the function of connecting with the sacred. Hindus believe (Eck, 1985) that *puja* connects the minds of the devotees to the Divine incarnated in the *murti* (image/idol). *Darshan*, 'seeing the Divine', is central to *puja*. Devotees believe that in *puja*, the eyes of the Divine meet those of the devotees and this exchange of glances between the two (through their respective eyes) provides the most coveted religious experience of the vision of the Divine for the devotee. It is further believed that this experience transforms the mind of the devotee (Eck, 1985). What *puja* does is transform world-centred consciousness into a God-centered consciousness. In this sense, *puja* is an individual encounter with the Divine regardless of whether it is performed with other devotees or not.

While language (of prayers, scriptures and religious music) plays a role in the restructuring of consciousness (from the self-centered to the God-centered), its use is not mandatory. Silence can be maintained throughout the *puja*. Since *satsang* is invariably a dialogue of the religious master/Guru and the disciples, language is essential. The disciples gather in *satsang* to learn from the master through discussions on the topics related to material as well as spiritual life and the connection between the two. Language is the most essential aspect of *satsang*. The structure of the discourse with the Guru in *satsang* is not predetermined. It is organised depending on the questions that devotees ask and the answers that the Guru provides. On the other hand, the structure of the language used in the *puja* (along with other actions mentioned above) is conventionalised and to a large extent predetermined. It is believed that the 'performative function of *puja*' is accomplished when it is performed according to the predetermined conventions. Therefore, any changes in the ingredients of *puja* are not really acceptable since the 'performative' force (or efficacy) of the ritual depends on its 'flawless performance' which would not admit change in linguistic code. However, in my earlier work (Pandharipande, 2006, 2010) I have shown that in the US diaspora, the English language is currently being used in worship rituals. Thus, the performative function of language depends on its structure dictated by conventions while the structure of the dialogue in the *satsang* is flexible. While language plays the role of a communicator/mediator in the *satsang* to facilitate communication between the Guru and the seeker/disciple/aspirant, it is the medium which helps the devotee to directly experience connection with the Divine. Thus, the primary function of language in *satsang* is informational while in *puja* it is performative. In *satsang*, language *informs* while in *puja*, it *performs*.

The last decade has seen a marked use of the internet in rituals including worship/*puja*. Globalisation is one of the major reasons for the migration of many Hindus to the US. There are over 200 Hindu temples devoted to numerous deities in the US. Hindus in the US have arrived from many countries such as India, Sri Lanka, Nepal, the UK, Malaysia, Africa, Fiji, etc. They cannot visit India (or whichever their native country may be) as often as they would like, so virtual/online *puja* offers a much needed opportunity for them to visit the temples of their choice in India and perform *puja* for the deities they wish ('Online Puja', n.d.). A number of major temples in India have opted for providing online worship of the deities such as Vishwanath, Durga, Hanuman, Vishnu, Shiva, and more. Online *puja* provides easy access to temples without the need to physically travel from the US to connect with the Divine.

The typical online *puja* does not have the restrictions of time or space and the need for a priest. Additionally, the materials (incense, flowers, food, etc.) are provided in their 'virtual' form which takes away the need for preparation time and material required in offline *puja*. The instructions for conducting *puja* are provided by the 'digitised voice' of the computer. There are many online *puja* websites currently available. For those who know the process, and do not need any instructions for *puja*, the computer-initiated instructions are superfluous. For example, Spiritualpuja.com ('Online Puja', n.d.) provides information about what *puja* is, its purpose and its role in the life of Hindus. It also explains every step of the ritual and its purpose. The instructions include: (a) choose temple; (b) enter temple; (c) click on the temple bell; (d) click on the incense; (e) click on the lamp to light up the flame; (f) offer food by clicking on the sweets. The religious music appropriate for the 'chosen' deity plays in the background.

Language in the online *puja* varies according to the context. For example, the instructions for the performer of the *puja* are typically in English. The choice of the language(s) of the chanting, recitation of *mantras*, prayers, etc., is determined by the type of *puja* scripture/sacred text. For example, the traditional classical *homam/havan*, 'fire sacrifice', is performed using Sanskrit while the recitation of *Hanuman Chalisa*, '40 verses dedicated to the deity Hanuman' and the *Satyanarayana Katha*, the ritual of reading the story of *Satyanarayana*, are typically in regional languages. Moreover, many websites allow translation of the prayers in the languages mentioned via a 'drop down' menu.

The difference between online and offline *puja*

The following discussion points out the differences between online and offline *puja*:

(a) The major difference between the two is that, at present, only a limited number of *pujas* can be performed online. For example, temple worship can be conducted online while the worship ritual which is performed at home with relatives, friends, etc., cannot be performed online since the participants vary and the household rituals vary according to the family traditions. In other words, 'institutionalised worship' according to the traditions sanctioned by the scriptures is currently conducted online. Therefore, the relatively more conventionalised rituals alone are performed online. Moreover, certain rituals such as weddings, naming ceremonies for children, house warmings and funerals among others, cannot be performed online because these rituals require the actual presence of the participants in the *puja*.

(b) As online *puja* is on the increase, there is a 'standardisation' of *puja* which does not admit variation of language, worship material and the steps in the ritual. In reality, offline worship ritual allows a great deal of flexibility in choosing languages and the types of flowers and sweets (for offering to the deity). The pattern of language use is 'standardised' in online *puja* while it is flexible in offline *puja*. The exception to this is the recitation of scriptures where the language of the scriptures cannot be changed. Thus, the online *puja* does not admit dialectal variation in the performance of worship. In other words, one can claim that the online *puja* creates linguistic hegemony (embedded in the overall structure of the worship ritual). Just as there is standardisation of ingredients in the online *puja*, there is linguistic standardisation as well which does not allow innovation, and individual modification in linguistic form as well as patterns of language use.

(c) Online *puja* is accessible to all, regardless of their caste, religion, age, language and type of temple. In this sense, online *pujas* are more liberal compared to their offline counterparts. In contrast to this, not all temples in India allow entry to non-Hindus (for example, the Padmanabha Temple Trivendram in Kerala and the Vishvanath Temple in Varanasi).

(d) Offline *puja* is not free from time restrictions. There are fixed days and times for *pujas*. An auspicious time is an important belief in Hinduism. On the other hand, online *puja* can be performed at any time according to the convenience of the performer of the ritual. However, many Hindus in the US (three out of eight whom I interviewed) try to perform their *pujas* at the same time as they are performed in India.

(e) Commodification of ritual is seen in online *puja*. Websites advertise the cost of performing online *puja*, the 'economy packages' for the *puja*. Additionally, one cannot proceed to perform *puja* unless a certain amount is paid upfront by giving a credit card number. While in offline *puja*, devotees can offer some money to the temple, minister, etc., this is voluntary and a worshipper can also choose not to give any money. In many temples, money is for the priest's services and for the maintenance of the temple. However, generally, no one is denied access to worship in a temple on the basis of their donation (or lack of it). Some websites do allow *puja* without payment. However, it is becoming more common for online worshippers to pay for the *puja*. This shows that while online *puja* is more democratic as it is accessible to all, it also creates a divide between those who can pay and those who cannot. This is a good example of how religion is being influenced by the market economy.

(f) Online *puja* does not take place in the same atmosphere as offline *puja*. The smell of incense, offerings of food, fruit, flowers, etc., people's movements back and forth in preparation for the ritual, the mixing of sounds of chants, temple bells, *bhajans* sung by the priest and the participants, the visual scene of colourful clothes, turmeric, *sindoor* and so on create an ambiance conducive for the ritual and the mindset oriented toward the worship ritual. A majority of people (9 out of 10) said that ambiance is important for the success of the *puja*. Ambiance is missing in online *puja*. If the goal of the worship ritual is to transform the mind, then it is important to note the difference between the two.

Discussion

In what follows, I will summarise the above discussion, present scholars' views on Hindu rituals and attempt to provide some answers. The above discussion points out that there is a difference between online *satsang* and *puja* and their respective offline counterparts. Also, there is a concern regarding the authenticity of online rituals, i.e. online rituals are not viewed by many Hindus (including priests/saints) as legitimate religious rituals. Some scholars (Fuller, 1992; Scheifinger, 2012) have claimed that the lack of presence of full paraphernalia in online rituals (compared to their offline counterparts) should not be an issue. Scheifinger (2012) claims,

> ...the practice of *darshan* – the key feature of *puja* – can be successfully mediated via the Internet. While the other senses may not be stimulated in the way that they are when conducting the *puja* in the offline setting, the importance of the sense of sight remains in an online *puja*. In both cases, a devotee is able to see and be seen by the deity. Therefore, in this crucial respect, the *puja* is not altered radically in the online form. (Scheifinger, 2012: 126)

Here, Scheifinger makes a distinction between 'core' and other features of *puja* and marks the *darshan* 'seeing the Divine' as a 'core' feature of *puja* and argues that online *puja* ritual is fine as long as 'core' features are maintained while some others are not. Fuller (1992: 246) supports Scheifinger's claim that abbreviations do not interfere with the legitimacy or efficacy of the rituals. Another issue concerning the purity of the online *puja* ritual has been discussed by Karapanagiotis (2010). She claims that the objection to cyberspace by many devotees due to uncleanliness (since computers can be used for 'impure' secular activities such as work and pornographic films) is not correct. Karapanagiotis (quoted in Scheifinger, 2012: 125) suggests two strategies to keep impurities out of the computer: firstly, keep the physical space around the computer clean and secondly, approach the ritual with a pure heart which must overcome the impurity. Scheifinger (2012: 125) further claims that Hindu goddesses and gods are not affected by perceived impurity and the environment may become impure by the presence of an impure person.

The question of the necessity of ambiance (mentioned in the earlier section) needs further discussion. The presence and absence of ambiance in offline and online *puja*, respectively, raises a question: on the one hand, we can say that online ritual is inadequate in providing the appropriate ambiance, while on the other hand, we can ask whether ambiance is a necessary ingredient to achieve the goal of *puja*. Scheifinger (2012: 126) correctly points out that not all props are necessarily present in offline *puja* and,

> ...indeed in Hinduism, acceptance of the idea that a *puja* ritual is symbolic is made explicit through the existence of a type of *puja* in which 'the physical form (of the deity) is carefully mentally reconstructed with such rituals as libations and flower offerings being exactly performed in the virtual reality inside the head'. (Smith, 2003: 144)

Any abbreviation associated with the practicing of online *puja* is also unproblematic. It does not interfere with the ritual's efficacy because ritual abbreviation and simplification are ubiquitous procedures that are allowed by the (Hindu sacred) texts themselves. It is observed that priests perform *puja* in different parts of India and take the liberty to use the fruits and flowers available in that part of the country. Moreover, as Scheifinger points out, the 'script' of *puja* does not necessarily require all props to be present each time *puja* is performed. It is certainly the case that abbreviation and variation of the ingredients of *puja* (texts, prayers, devotional songs, length of readings, etc.) are acceptable to most people (devotees and saints included). In this context, it is important to ask why some devotees, saints and priests (with whom I had conversations) do not accept this view about the legitimacy of online *puja*. They claim there is a difference between face-to-face *satsang* and *puja* of the *murtis* (deities) in a temple and their online counterparts. Their view is that the media-mediated connection is the same as the connection with a photograph. Just as seeing a photograph is not meeting the person, seeing the virtual image of the Guru in *satsang* and seeing a deity's virtual picture in *puja* are not equivalent to their respective offline encounters with the Guru and the deities.

Another concern about the efficacy (or lack of it) was brought to my attention when I met with a saint and asked him about the legitimacy of online *puja*. He said,

> ...online *puja* is not legitimate because the images of Gods are not 'real *vigrahas (legitimate deities)*' because the idols acquire 'sacred status' only when consecrated in the temples by the priest or they are 'authenticated by the saints'. In the online *pujas*, they are merely simulated images of the deities. They cannot have efficacy of the real deities...

In this context, it will be useful to examine the history of Hinduism to show that the change of method/media for expressing Hindu beliefs is not unique to the present context. Historical developments in Hinduism provide evidence for changes in the medium of expressing Hinduism. Oral tradition was the exclusive method of expression and transmission of Hinduism and the gradual acceptance of the change in the tradition. The oral traditions in the ancient time of the Vedas continued until they were written down for the first time in the 6th century BCE (Avari, 2007: 69–70). It is important to note that oral transmission later, in the 18th century, was supplemented by print media. In contemporary India, oral transmission of scriptures and rituals continues along with the reading of sacred texts such as the Vedas, Upanishads, the epics (the Ramayana and the Mahabharata) and hundreds of other scriptures dedicated to many deities (local and pan Indian). At present, no question is raised about the authenticity and efficacy of the written scriptures and their use in the *satsang* and/or *puja* ritual. Similarly, the change in the language used for *satsang/puja* is observed in the history of Hinduism. While Sanskrit enjoyed the status of the language of religion *par excellence* (and continues to do so today), other regional languages of India began to be used in religious rituals from the 13th century. The use of English as a language of Hinduism and its use in the *satsang* are well attested since the 19th century when saints such as Vivekananda used English for communication. Twentieth-century India and the US show the use of English for *satsang*. Additionally, it should be noted that while English is used in wedding/*puja* rituals in the US (Pandharipande, 2010), its use in *puja* rituals in India is not widespread. The important point to note here is that change in the media of communication of Hinduism is not a new 21st-century phenomenon. Therefore, digital media as a medium for communicating religion in general, and for *satsang* and *puja* in particular, should not be problematic. The history of Hinduism also shows that the changes were met with resistance but were gradually assimilated and accepted within the tradition. What we observe here is that the 'script' of rituals (language in particular) has undergone many changes in the history of Hinduism and the changes have been absorbed into the system as 'legitimate'.

The questions that must be asked in this context are

(a)   What is the mechanism of the conventionalisation or acceptance of the change in Hinduism? In other words, how is the change authenticated?
(b)   Why is the change at first rejected and then accepted?

In my work on language change in Hindu rituals in the US (Pandharipande, 2010, 2013, 2015), I have claimed that the use of English for Hindu rituals is authenticated by the saints in Hinduism (in India and the US). Unlike Christianity, Hinduism does not have the institution of the church which authenticates or rejects changes. From time immemorial, changes in rituals and languages have been authenticated and thereafter conventionalised in Hinduism by the saints who are perceived as the highest authority. For example, Prabhupada, founder of the 20th-century ISKCON (founded in the US) recognised the reading of the English translation of the Bhagawadgita as a legitimate ritual on par with its reading in Sanskrit, the original language of the text. Similarly in the 13th century, Jnaneshwar, a saint in Maharashtra, used Marathi for Hindu rituals. Therefore, I argue that if the saints approve the communication of religion through media, its use is authenticated. Currently, most of the Hindu saints who visit the US use electronic media for their *satsang*, thereby legitimising it. In contrast to this, at present, the saints do not use electronic media for performing *puja*. Therefore, media-mediated communication is better accepted by the devotees in *satsang* as opposed to *puja*.

Signifier and signified: A changing equation

The question why a change in the medium of communication of religion is accepted or rejected needs a deeper discussion from the perspective of religious studies. The answer lies in the analysis of the process of conventionalisation of rituals and the change they undergo at different points in time and space. Religious rituals and their structural and functional complexity have been discussed in a large body of research (Helland, 2012; Leach, 1968; Malinowski, 1954; Turner, 1969, among others as discussed above). Despite their differences, scholars agree that religious ritual is a process to connect with the Divine and that the process/action is carried out through symbols which are conventionalised in and through the religious tradition. For example, in Hindu *puja* ritual, the symbols include the representation of deities in the *murti/vigraha* 'statues' (literally, concrete symbols). The religious offerings such as flowers, incense, food, oil lamps, prayers, the recitation of sacred texts, etc., participate in the ritual activity. The ritual objects, including the deities, individually and collectively, contribute toward establishing a connection with the Divine. The devotees believe that these symbols function as signifiers which signify the appropriate context, and the process which leads to the connection with the Divine. The major argument for the perceived relatively lower efficacy/validity of online rituals compared to their offline counterparts is because the online rituals have not yet established the signifier–signified relationship between virtual and physical images. The ritual meaning is conveyed or realised by the conventionalised relationship between the signifier and signified. For example, a particular statue of a deity, Shiva for example, is generally ritually established in a temple. The ritual installation of a statue in a temple is viewed as *pranapratistha* (literally, infusing life in the statue to make it 'Divine'). The Hindu belief is that the rite/ritual brings the power/life of the deity into the statue/icon (for further discussion, see Bharne & Krusche, 2012). Moreover, it is believed that worship should not be offered to the *murti* (the statue) in a temple unless the rite of *pranapratistha* has been performed. This rite is done by a priest with ritual chanting of the appropriate *mantras* (sacred verses believed to have the power/efficacy). In other words, the signifier (statue) functions as the signified (powerful deity) only if the rite has been performed. Similarly, for *satsang* to achieve its goal of transmission of knowledge through a dialogue between the Guru and disciples, it is important that the Guru is present in person. I argue that the relationship between the signifier and signified is established or conventionalised within a culture of a religious community through tradition. Language is an integral part of the whole equation of the signifier and signified. That is, language, along with ritual objects, process and the physical context are culturally conventionalised signifiers for the intended signified meaning/goal of the ritual (*puja*, for example).

Additionally, it is important to note that the above-mentioned objects, actions and language are authenticated in and by tradition and, therefore, in order to ensure efficacy of the ritual, each of these must be exactly 'as prescribed' in the tradition. Thus, every aspect of *puja* ritual, time, place, agent and language is irreplaceable. For example, the place of the *puja* must be clean, fresh flowers and fruit must be offered, the performer of *puja* must bathe before performing the ritual and certain texts must be read for certain *pujas* (for example, *Hanuman Chalisa* for the deity Hanuman, and *Durga Saptashati* for the goddess *Durga*). The sequence of actions is also important. The oil lamp must be lit before starting the ritual and the fruit/food must be offered after reciting scriptures. The major point I want to make here is that the signifier–signified relationship is conventionally established and therefore, when some of the signifiers are absent, the devotees do not perceive the *puja* to be authentic. For the worship ritual to function as the signifier of the connection with the Divine, every part of it must be 'as it is supposed to be' according to the conventions of the tradition. In other words, at any point in time, the devotees believe that the 'script' for the rituals must be followed flawlessly without admitting any change. Since online *puja* does not have ritually 'consecrated' deities, many do not accept online *puja* to have the same efficacy as offline *puja*. Scholars have agreed on the fact that the structure and function of religious rituals are not frozen, rather they are dynamic and change with the socio-religious context. According to Saussure (Holdcroft, 1991), as already mentioned earlier, the relationship between the signifier (the form of a sign) and the signified (its meaning) is arbitrary or conventionally determined. In *puja* ritual, all of the physical objects, including the language and the *murti* (statue), are part of the larger signifier, *puja* ritual, which signifies the connection with the Divine.

As mentioned in the earlier section, it is important to remember that Hinduism has gone through many changes in its history. In particular, the relationship between the signifier and signified has been changing through the passage of time. New languages and new objects (e.g. apples, or other fruits in place of coconuts) were introduced into the rituals and new spaces were accepted as appropriate for *puja*. In fact, the ancient ritual of fire sacrifices was not performed in any temple; rather, it was performed in temporary structures built for Vedic sacrifices which were ritually destroyed when the ritual was concluded. Another important change to note is that Vedic Sanskrit was replaced by classical Sanskrit and later by regional Indian languages. In other words, the signifiers at the micro level were changed and new substitutes took over their ritual function of signification. It is also important to note that the 'mediated communication/expression' of religious meaning is not new to Hinduism where the 'concrete deities (with different forms)' were introduced in the system which symbolised/signified abstract meanings (for example, Vishnu as the sustainer of the world, Shiva as a destroyer, Goddess as feminine power, etc.). The *puranas* are full of symbols and deities which project a symbolic universe quite different (not necessarily unrelated) from its Vedic counterpart. Similarly, the ancient tradition of spreading religion orally was supplemented by the written word in the history of Hinduism, where text became a signifier of meaning, like its 'oral' counterpart. Thus, when the process of offline *puja* is replaced by its online counterpart, the simulated images of the actual objects function as signifiers and signify the ritual object at the micro level and *puja* at the macro level. Since the phenomenon of online *puja* is relatively recent, the relationship between the signifier and signified is not yet fully conventionalised in the tradition. In contrast to this, online *satsang* is better accepted and has not been met with resistance. In a secular context, televised lectures, conferences, meetings and concerts have become quite routine and their legitimacy is acknowledged. The signifier–signified relationship between the digital image of the Guru (or the person who conducts the *satsang*) has been conventionalised in the secular as well as the religious domain over a long period of time. However, the relationship between the signifier (the ritual objects) and signified (the connection with the Divine) in *puja* is not yet conventionalised in Hindu tradition. If history is any indication, the change in the method of conducting *puja* will eventually be accepted as an authentic practice in Hinduism.

Lundby (2012), while discussing theoretical approaches to the study of digital religion, correctly points out the need for analysing the religion and media studies approaches to understand the underpinnings (the goals and processes) involved in the production of mediated religion. He notes,

> However, none of the aspects of religious studies can prosper without further insight into how media and communication processes work in contemporary society. Religious studies could learn from media studies in order to undertake proper research into religion and media. Similarly, media studies could approach religion in the same way as any symbolic field in the media. However, a deeper analysis requires greater understanding of religious traditions and their symbolic universe. (Lunby, 2012: 225–226)

In the context of Lundby's view, the question about the efficacy of virtual symbols in online rituals created in the media can be discussed. I would like to claim that online religious rituals stand at the intersection of religion and modern computer (internet) media. The criteria for the authentication of virtual symbols would differ from the criteria for their authentication in religion. For example, internet media can produce perfect virtual images of Hindu deities. However, in order for them to function as symbols or signifiers of the deities, the religious authority would have to authenticate them and the relationship between the images and (signifiers) and signified deities would have to be conventionalised in the tradition. Since online rituals are a relatively recent phenomenon, the process of conventionalisation and acceptance of the ritual virtual images as symbols is not fully established yet. This is the reason why not all devotees and saints believe in the efficacy or legitimacy of the virtual images in online rituals.

Conclusion

In conclusion, I would like to summarise the above discussion as follows: (a) Online *satsang* and online *puja* differ from their respective offline counterparts because they are both performed through the mediation of technology. (b) The interaction between the participants and the Guru in *satsang*, and between the devotees and the deities is mediated by technology. Thus in both, the signifiers (Guru in person in *satsang* and statues of deities in *puja*) are replaced by their digital images in the online context. The referents (Guru in *satsang* and deities in *puja*) are the same as their respective offline counterparts. (c) The discussion points out that the reason for a 'perceived' lack of authenticity of online rituals is because people/devotees do not accept digital images as legitimate signifiers of the signified religious meaning. Interaction with 'images' is not equal to interaction with the actual person (in *satsang*) or the deity (in *puja*). (d) Finally, I claim that every part of the ritual acts as a signifier for the signified religious meaning and that the relationship (signifier/signified) is established by religious conventions.

The history of Hinduism shows that its conventions can change according to the social contexts and that new signifiers are introduced into the system to signify the same meaning (e.g. change in language of religious rituals). The new equation of signifier–signified is authenticated by the saints since the saints themselves conduct online *satsang* and thereby endorse the relationship between the new mediated virtual images and their ritual meaning. Finally, I would like to raise some questions in this context: (a) whether the change in the relationship between the signifier and signified changes the religious meaning; (b) how the change in the signifiers at the micro level (for example, the virtual image of flowers in online *puja*) impacts the overall signification of the *puja* ritual at the macro level; and (c) whether virtual online *puja* can be accompanied by or mixed with its offline counterpart. Since the phenomenon of digital religion is widespread, these questions are relevant in the context of other religions as well.

References

Avari, B. (2007) *India: The Ancient Past*. London: Routledge.

Bharne, V. and Krusche, K. (2012) *Rediscovering the Hindu Temple: A Sacred Architecture and Urbanism of India*. Newcastle: Cambridge Scholars Publishing.

Durkheim, E. (1995) *The Elementary Forms of Religious Life* (K. Fields, trans). New York: Free Press. (Original work published 1912.)

Eck, D.L. (1985) *Darshan: Seeing the Divine Image in India*. Chambersberg: Anima Books.

Free Online Satsang (n.d.) I-CODA Free Online Satsang: Live Interactive Meeting with Swami Premodaya. See www.meetup.com/Swami-Premodaya-satsang-meditation-spiritual-guru-advaita/events/186969872 (accessed February 2017).

Frisk, L. (2002) The satsang network. *Nova Religio: The Journal of Alternative and Emergent Religions* 6 (1), 64–85.

Fuller, C.J. (1992) *The Camphor Flame: Popular Hinduism and Society in India*. Princeton, NJ: Princeton University Press.

Geertz, C. (1973) *The Interpretation of Cultures*. New York: Basic Books.

Helland, C. (2013) Ritual. In H.A. Campbell (ed.) *Digital Religion: Understanding Religious Practice in New Media Worlds* (pp. 25–40). London: Routledge.

Holdcroft, D. (1991) *Saussure: Signs, System, and Arbitrariness*. New York: Cambridge University Press.

Karapanagiotis, N. (2010) Vaishnava cyber-puja: Problems of purity and novel ritual solutions. *Heidelberg Journal of Religion on the Internet* 4 (1), 179–195.

Kreinath, J., Hartung, C. and Deschner, A. (eds) (2004) *The Dynamics of Changing Rituals: The Transformation of Religious Rituals Within Their Social and Cultural Context*. New York: Peter Lang.

Leach, E. (1968) Ritual. In S. Hugh-Jones and J. Laidlaw (eds) *The Essential Edmond Leach*. (pp. 165–73). New Haven, CT: Yale University Press.

Lundby, K. (2012) Theoretical frameworks for approaching religion and new media. In H.A. Campbell (ed.) *Digital Religion: Understanding Religious Practice in New Media Worlds* (pp. 225–237). London: Routledge.

Malinowski, B. (1954) *Magic, Science and Religion and Other Essays*. Garden City, NY: Doubleday.

McGuire, M. (1997) *Religion: The Social Context*. Belmont, CA: Wadsworth Publishing.

Niranjani (n.d.) Niranjani: View on life, spirituality & meditation. See niranjani.wordpress.com/tag/satsang/ (accessed February 2017).

Online Puja (n.d.) Online puja. See www.spiritualpuja.com/index2.html (accessed February 2017).

Pandharipande, R.V. (2006) Ideology, authority and language choice: Language of religion in south Asia. In T. Omoniyi and J.A. Fishman (eds) *Explorations in the Sociology of Language and Religion* (pp. 141–164). Amsterdam: John Benjamins Publishing Company.

Pandharipande, R.V. (2010) Authenticating a tradition in transition language of Hinduism in the US. In T. Ominiyi (ed.) *The Sociology of Language of Religion: Change, Conflict and Accommodation* (pp. 58–83). London: Palgrave Macmillan.

Pandharipande, R.V. (2013) Does Religion Promote as well as Retard Language Maintenance in a Multilingual Context?: The case of Hinduism. Paper presented at the International Conference on Sociology of Language of Religion, New York University, July 17–19.

Scheifinger, H. (2012) Hindu worship online and offline. In H.A. Campbell (ed.) *Digital Religion: Understanding Religious Practice in New Media Worlds* (pp. 121–127). London: Routledge.

Smith, D. (2003) *Hinduism and Modernity*. Oxford: Blackwell.

Turner, V. (1969) *Structure and Antistructure*. London: Routledge.

10 Virtual Allegiance: Online 'Bay'ah' Practices within a Worldwide Sufi Order

Andrey Rosowsky

Employing a multimodal-inflected discourse analysis, this chapter revisits the fluid and transformative space that exists in the intersection between orality and literacy. Using the context of online ritual practices, in particular, instances of online initiation into Sufi fraternities, it presents data that suggest that the instantiation of originally spoken practices into online contexts results in a range of language modes which, on the one hand, try to emulate spoken conventions, and, on the other, transform the spoken word into different forms and modes of literacy. Simultaneously, these modes and forms can make for more efficient access to ritual practices and equally serve as new gatekeepers to what have hitherto been accessible cultural practices. The results of the analysis contribute to the 'literacy debate' inasmuch as they provide yet another example of how the written language can often serve to obfuscate and deter rather than make transparent and liberate. It is also possible that what we are witnessing here is a significant historical factor in how language is theorised in the new digital age, akin to the impact on language caused by the introduction of print technology 500 years ago.

Introduction

> A generalised dichotomy between oral and written texts has much power, with the former being linked to interpersonal concerns, informality in style and dialogicity and the latter being linked to content focus, greater formality and monologicity. (Ong, 1982)

The practice of highlighting the stark contrast between orality and literacy, exemplified in Ong's comment is, despite the earlier work of Parry (1971), Lord (1960), McLuhan (1962) and others (Eigenbrod, 1995), still to be found in much that is written on the orality–literacy conundrum, particularly in the field of education (Purcell-Gates, 1998; Soukop, 2007).

Yet, Ong's work and the earlier work of others (Lord, 1960; Parry, 1971) convincingly persuade us that such a contrast is never as stark as imagined and, with the advent of electronic communication, and what has been called 'secondary orality', the relationship between these two communicative modes is even more complex and dynamic. This chapter, therefore, presents an example of how newer instantiations of this complexity and dynamism are enacted through ritual in online digital spaces and presents data that suggest the instantiation of originally spoken practices into online contexts results in a range of language modes which, on the one hand, try to emulate spoken conventions, and, on the other, transform the spoken word into different forms and modes of literacy.

The 'Great Divide' debate in the field of literacy studies has often had a close link to faith and religious practice. Scribner and Cole's (1981) study of Qur'anic literacy (among other literacies), along with Street's (1985) influential *Literacy in Theory and Practice*, in some respects, satisfactorily challenged the once prevailing view of literacy as not only a 'breakthrough' technology which served as a means to advance societies but also as inherently capable of shaping thinking, leading to the development of logic, transparency and rationality (Goody, 1968, 1977; Havelock, 1976; Ong, 1982). Such a view still has more contemporary guises in the work of Baron (2008) on online language practices which, incidentally, seldom refers to studies, directly or indirectly, within the field of 'New Literacy Studies'. This may say, however, more about the distrust between disciplinary silos than it does necessarily about opposing views and argument.

It is noteworthy that both the seminal studies mentioned (Scribner & Cole, 1981; Street, 1985) had a form of religious (Qur'anic) literacy at their heart. In Street's follow-up volume in 1993, other forms of religious literacies, again including Qur'anic ones, also featured regularly (Bledsoe & Robey, 1995; Probst, 1995; Reder & Reed, 1995). Boyarin's (1995) volume on the ethnography of reading, likewise, presents a number of studies foregrounding the literacy practices of religious rituals (Baker, 1995; Digges & Rappaport, 1995; Noakes, 1995). Exposing the power of the written word to both obfuscate and de-democratise was evident in many of these studies, particularly, but not exclusively, in the ways that ritual practices were co-opted by authority and ruling elites to reinforce inequalities and social differentiation. Therefore, within the tradition of literacy studies as they developed in the last decades of the 20th century, the literacy practices of religious rituals provided much of the exemplification for the arguments that literacy is a social practice contingent culturally, socially and historically. Furthermore, such studies and others (see Graff, 1979) challenged the view that vernacularisation, and literacy as its technological handmaiden, encouraged certain cognitive affordances such as objectivity and context-independent thinking and certain social advantages such as democratisation and individualism.

Online religion is to do with practices that can be carried out online – either the virtual opportunity to enact rituals or even the possibility to forge new, or at least transformed, online religious practices (Helland, 2007). Furthermore, Helland (2007) makes the useful distinction between 'religion online' and 'online religion'. The former represents those sites that serve as depositories of information and knowledge about different religions and tends to have a unidirectional intention from 'one' to the 'many'. In particular, in religion online the environment was (and remains) biased toward *textual* representations of religious beliefs.

By contrast, and driven by the capabilities of Web 2.0 technology, online religion refers to sites that allow for more interaction, more accessibility, more immediacy and are more 'many to many' in orientation. The Vatican website, The Holy See, is an example of a 'one to many' direction but there are an increasing number of sites that allow for, for example, virtual worship, virtual pilgrimage and a sense of belonging to a virtual community. This different dynamic has obvious implications for questions of authority with online religion spaces having the potential for flatter hierarchies vis-à-vis their participants.

More recently, the term 'digital religion' has been used to denote 'religion that is constituted in new ways through digital media and cultures', which could lead to 'a new understanding of religion' (Campbell, 2013: 3). I would suggest further that an exploration of digital religion and digital spaces may also lead to new understandings of language. The advent of print technology and the proliferation of reading practices beyond the privileged cloisters of monastic scriptoria and medieval universities (Febvre & Martin, 1976) meant that the unity of sign and signifier of the pre-Reformation age was soon followed by a separation of form and meaning leading to an emphasis on the referential function of words and language (McLuhan, 1962). It could be that the language of digital religion either reverts to that previous unity of sign and signifier, and there is evidence that this is already happening (online calls to prayer for example), or that the referential function could be consolidated but expanded to include other semiotic resources, or, perhaps more excitingly, transformed into a newer conceptualisation of language as yet unformulated and invisible. This chapter presents data that could meet all three of these possibilities.

Recent technological advances afforded by the internet provide us with another, more immediate, context for examining the complex set of interweaving discursive practices operating within the space between orality and literacy. Of particular relevance to this chapter is what O'Leary (1996), drawing on Ong, has to say (a) about the nature of aspects of language change in those historical religious contexts that accompanied the growth of print technology and (b) what this might have to say about language and faith in the digital age – a moment possibly every much as significant for human culture and civilisation as the latter.

The traditional divisions of religious life into the profane and the sacred, as originally defined by Durkheim (1912/1976), have their counterparts in the terms 'sacred cyberspace' and 'profane cyberspace' (Jacobs, 2007). Thus, O'Leary (1996) in his seminal article on sacred cyberspace drew a useful comparison between Ong's (1982) elaborately developed correlation between societal literacy (and especially print literacy) and the shift to more literal or prosaic forms of Christianity at the time of the Reformation, and the possible impact on language of the present historical moment with its digital online spaces and their use as religion online and online religion. This line of thinking has, of course, antecedents in the prophetic work of McLuhan (1962) whose prescience anticipated the shift from literacy, and its unique sequential and visual nature, to a more holistic and multisensory return to orality via electronic means – what Ong (1982: 136) calls 'secondary orality'.

More generally, these incipient musings of O'Leary on the relationship between the internet and faith have now developed into an increasingly significant body of work focusing on this nexus. Campbell (2007, 2010, 2012, 2013) and others (Casey, 2006; Jacobs, 2007; MacWilliams, 2006; Radde-Antweiler, 2006) have documented and further explored a range of faith contexts and digital platforms in their endeavour to theorise what is happening in the present moment with faith and digital spaces. Little of this ground-breaking work to date has focused on language despite O'Leary suggesting some foundations for a linguistic exploration. This chapter, along with others in this volume, seeks to contribute to the developing academic study of the internet and matters of faith and language.

At the heart of any consideration of religious ritual practice online is the extent to which a ritual in its online manifestation has undergone an alteration from the corresponding ritual carried out in its traditional context (Pandharipande, this volume; Scheifinger, 2013). From a language point of view, despite the capability of digital spaces for multimodal communication, the movement from substantially oral rituals to online spaces inevitably draws on the written word in ways that are absent, concealed or unnecessary in offline spaces.

Language, Religion and Technology

That this historical moment in the theorising of language could be as significant as the Reformation is perhaps suggested by the sheer amount of online activity related to religion and faith practices. Indeed, it is important to recognise that the traditional association of technology and modernity with secularisation is challenged by the massive presence on the internet of religion. Helland (2007: 957), for instance, reports that already by 1996 'there were three times as many sites concerning God and spirituality than there were concerning sex' and that by 2000 'more people were using the

Internet for religion and spiritual purposes than were using the medium for online banking or online dating services'. Moreover, the online presence of religion is so predominant that Helland (2007: 957–958) reports that within the core titles used by search engines to categorise websites the second-largest category on the entire World Wide Web is 'Religion and Spirituality' with slightly more sites than each of the categories: 'Science', 'Sports' or 'Shopping'.

A key linguistic observation that O'Leary makes, and which was also partially foreseen by McLuhan, though not in a directly religious context, was the shift from a holistic and integrated appreciation of the 'word' to a literal and more referential, meaning-centred focus. If in pre-Reformation Christianity and pre-colonial Hinduism (Yelle, 2013), the language of faith practices was primarily an oral mode – despite the presence of sacred texts whose reading was invariably an occupation of a priestly elite – the word and its referent were seen as interchangeable and indivisible – the word in such contexts was seen as an efficacious phenomenon in its own right with a power (magic) and a sanctity intrinsic to its existence. In this sense, all words, particularly those used in a religious context, were performative. O'Leary (1996: 788) quotes the words of the Latin Mass which, pre-print technology, had an efficacy born of, in language theory terms, the unity of sign and signifier. The wafer and wine became flesh and blood by being called into being by the authorised priest. The advent of print technology and the reforming transformation of the European church led to a divorce between sign and signifier – some of this was due to the vernacularisation of the liturgy and other church practices – and a change in the linguistic theory espoused (see also Pandharipande, this volume). From now on, the Eucharist and its language would be symbolic rather than a literal act of transubstantiation. The language used would be primarily referential and this split between a word and its referent has led many, in keeping with the spirit of Weber's notion of disenchantment, to claim it as a key driver in the development of modernity itself (McLuhan, 1962; Ong, 1982; Yelle, 2013).

Ong and O'Leary reference the European Reformation to exemplify the orality literacy dichotomy. Yelle (2013), on the other hand, reveals key parallels between the Reformation, its impact on language and the consequences of the colonial violence in India. Drawing a comparison between pre-Reformation Christianity and pre-colonial Hinduism, Yelle is able to show how the move from orality to literacy, promoted politically and culturally by the colonial powers, Britain in particular, had significant implications for the future development of Hinduism in India up to and beyond independence. The denigration of oral practices such as mantra chanting were equated by orientalist scholars to the recitations of the rosary and other enchantment practices – these were the 'vain repetitions' used by many Protestant scholars (Yelle, 2013: 105–107) to attack many practices of the Roman church. The same weapon was used by orientalist scholars to denigrate Hindu practices and, subsequently, through targeted selection and translation of sacred texts, to foreground literal and referential meaning in Hindu worship.

In both of these contexts, the arrival of print technology played a significant role with a staged vernacularisation of sacred language practices encouraged alongside a literal interpretation of sacred texts to the detriment of the efficacy of the language of ritual performance. With such profound and lasting changes to the language of such faith practices caused/facilitated by a technological shift, it is not unlikely, but perhaps still to be revealed, that a similar set of language changes could be accompanying the digital revolution.

The study of faith and language is an emerging sub-discipline drawing on the strengths of both its parents, the sociology of language and the sociology of religion. In Fishman's (2006: 18) *Decalogue of guiding principles for a sociology of language and religion*, mention is made of 'sources of sociocultural change [that] are also sources of change in the sociolinguistic repertoire for religion'. This chapter, like the other chapters in this volume, will refer to an obvious 'source of sociocultural change', and a sociocultural change in itself, the internet, to explore some of the sociolinguistic implications that arise from the enactment online of religious rituals hitherto carried out face to face, person to person and usually hand to hand.

These implications include (1) the continuing encroachment, and adaptation, of English into cultural domains associated more readily with the other major languages of Islam; (2) the uncertainties and inconsistencies apparent in the online convergence of different languages and scripts by website authors and readers; (3) some of the linguistic features of ritual enacted online.

Computer-mediated communication

Much research on online environments has focused on literacy practices often viewed through a multimodal lens recognising the range of modes online environments, such as websites, can offer their interlocutors (Androutsopoulos, 2006; see pages 7–9 of this volume for a more full discussion).

Written language in many online contexts, as has been acknowledged, often adopts a hybrid speech-writing mode. This is true for discussion forums, chat rooms as well as for texting and instant messaging (Baron, 2008; Crystal, 2006, 2008). It may be the case that all forms of technological innovation related to human communication have had this hybridising characteristic. The invention of the telephone (Gillen, 2000; Hopper, 1992) and the telegram (Standage, 1998) in the late 19th century, the film, the radio, the television, the mobile phone, email (Baron, 2003, 2008; Danet, 2001) have all developed hybrid modes of communication emphasising alternately speech or written patterns (Baron, 2013; Lee, 1996). The internet, therefore, offers a rich context for exploring that see-saw-like movement between orality and literacy which seems to be a regular companion of communication technology breakthroughs (McLuhan, 1962).

Performance and initiation

Bauman (1974; Bauman & Briggs, 1991) understands performance as language practice facilitated and constrained by a frame which determines the parameters that define performance vis-à-vis other verbal or linguistic modes and genres. For example, he posits a continuum where one pole represents performance which is characterised by verbatim recitation and enunciation of texts, sacred and other, with little, or at the very extreme, no room for veering away from the set script and no question of improvisation. As one moves along the continuum, space appears for less word-for-word performance and more room for improvised interjections and spontaneous insertions, while remaining within the frame of performance. The performance frame for a religious ritual such as the anointing of a new king would be fixed and time-honoured (thus explaining why it is 'fixed') with performers left with little or no opportunity for interpretation or embellishment. Somewhere in the middle of the continuum might be a traditionally scripted play where the words uttered may be unchangeable (although most Shakespeare performances allow for abridgement) but the possibility exists for interpretation through actorial voice, gesture and other aspects of body language. Towards the 'impro' end of the continuum, we might find storytellers who, given a loose scenario, can build a performance. Larry David's 'Curb Your Enthusiasm' TV series or Mike Leigh's films are known for their innovative improvised approach to acting but one is never in doubt that here is a performance and here are actors. In fact, one of the games that modern television writers of the mockumentary style play is to try and get the audience thinking or saying 'are they acting or is this real?' This is very much at the extreme end of the performance continuum. Of course, taking this metaphor further leads us to consider performance as underlying regular behaviour as we 'perform' our identities as understood by Butler (1997). For the purposes of this chapter however, performance has the specific definition of verbal performance in scripted and unscripted modes (Bauman & Briggs, 1991).

Ritual as performance has been researched in both sacred and secular spaces (Conquergood, 2013; Schechner, 1988, 2002; Turner, 1969). Some defining characteristics of sacred ritual are its multifaceted audience orientation which often includes a congregation or other social gathering, a transcendent being (or orientation) and, in this technological age, another audience made up of those who participate virtually, either via radio and television (initially) or via online channels. According to Bauman there is (a) an enhancement of expression and experience through performance even though more attention is paid to the formal and indexical properties of speech than to their referential or propositional meanings. It also draws attention (b) from the audience to the external and evaluative nature of performance.

These, then, are two crucial features of performance. Firstly, performance makes available to an audience the 'enhancement of experience',

> ... it is marked as available for the enhancement of experience, through the present enjoyment of the intrinsic qualities of the act of expression itself. (Bauman, 1974: 293)

There are notions here of striving for authenticity and accuracy, of the performer 'getting it right'. Secondly, performance is accountable and subject to the evaluation of an audience,

> Performance involves on the part of the performer an assumption of accountability to an audience for the way in which communication is carried out, above and beyond its referential content... (Bauman, 1974: 293)

Traditional and offline religious initiation is a performance that takes place within such a framework meeting the key characteristics of audience experience, audience evaluation, entextualisation of sacred scripts and emphasis on indexical and formal features of the speech act. Due to the regular occurrence of time-honoured phrases, formulas and exhortations, according to the Bauman continuum, much ritual initiation is situated towards the verbatim pole. Initiation performance can appear towards the improvisation pole in some contexts but even here there is more of an admixture of modes which allow for both word for word and improvised utterances (e.g. Wiccan initiation rites).

Always synchronous and physically embodied, such ritual partakes in the material, the emotional and the transcendental, leaving participants conscious of the experience they have just had – however assessed. Above all, this is an orality experience where words are uttered out loud or heard by those taking part. As Eliade (1965: 1) reminds us, *'Par initiation on comprend généralement un ensemble de rites et d'enseignments oraux'* ['By 'initiation' we generally mean a collection of oral rites and teachings']. Texts as artefacts may be present to assist recitation and enunciation (more common in some faith contexts than in others) but often utterances are entextualised via the oral medium through memorisation and transmission.

When such rituals go online, many aspects of the performance change or become problematic. It might even call into question the viability of maintaining 'performance' as a frame for understanding ritual online in a 'a sacramental space'.

Audience is suddenly invisible, unknown or even absent. Retaining the transcendent audience may be possible but only in the context of a decontextualised event compared with offline ritual. Some might suggest that the sanctity of the ritual is only established and maintained by time-honoured practices where recitation and audience are physically instantiated. The move online removes these guarantees and leaves the participant adrift in an imagined (virtual) ritual rather than a real one. It is also the case that there is a distinct move from the oral word to the written word with literacy demands now taking centre-stage, where previously they were less important and indeed could be avoided altogether. Nevertheless, as with other online platforms and contexts, conventions have arisen that seek to emulate or symbolise the oral conventions that cannot, strictly speaking, transfer in their original forms.

Computer-mediated communication, language and religion

This chapter draws on a range of approaches to bring attention to the 'entangled' (Bock, 2015) set of linguistic practices, many involving the written language, that accompany the transfer of ritual from offline to online contexts. The sociocultural premise that underpins this study recognises that language practice is always socially and culturally situated and that such a situatedness often draws on historical and geographical sources. Any discussion or analysis of ritual, however defined, takes place in the wider context of the relevant faith or ritualistic background within which the particular ritual takes place. A Latin Mass which takes place in Poland in the 21st century must be understood against the backdrop of a Catholicism that (a) pertains to the west of the Slavic world but not to the east; (b) has a history of language change from ecclesiastical Latin to vernacular languages prompted by Holy See decisions made at the beginning of the 1960s; and (c) has seen a counter-movement to reinstate ecclesiastical Latin in some Catholic contexts by the end of the 20th century. Why the Catholic Church uses Latin at all is another sociolinguistic story. And so on.

Therefore, a researcher brings to the analysis an interpretation that should be as full as possible while recognising the limitations of any one person to encompass and manifest the myriad strands of sociocultural relevance within any one particular item of analysis. In this chapter, which is a sociolinguistic analysis of online ritual, both micro- and macro-levels of analysis will be employed with references to lexis, morphology and syntax on the one hand and transnational organisations and their communications on the other.

There is also an ethnographic dimension to the study as I confess my personal association with some of the practices described and some of the broader contexts within which they are manifest. As a member of one of the Sufi associations mentioned in the analysis below, I am furnished with the insider-outsider perspective common to many ethnographic approaches. Having participated in some aspects of the practices outlined below, I am recognisant of the risks such proximity poses to the research process and, at the same, aware that such participation has given me a privileged stance vis-à-vis the research questions. My involvement in ritual practice and my subsequent reflection on that practice from a linguistic point of view can add unique depth to the analysis but, at the same time, the risk of observational blindness – we cannot always discern the strange in the familiar – is ever-present. Being able to operate across the languages involved in these ritual practices conversely can allow a sharpening of one's insight into their spiritual and emotional dimensions. For example, the issue of meaning and performance in ritual practices is very much opened up by reflecting on the language use within them.

Although this study makes primary use of a multimodal-inflected discourse analysis, it will be impossible not to draw attention to other theoretical and analytical frameworks in this process. As in the example above regarding Latin and Poland, word-level analysis will inevitably pose questions about or make connections to wider discursive channels and different modes of communication. The postmodern recognition of the pluralisation of literacy as 'literacies' which took place towards the end of the 20th century (Barton & Hamilton, 1998; Gee, 1990) acknowledges this multifaceted and multidimensional aspect of literacy practice and shares in the pluralisation meme recognised in multifaceted concepts such as 'identities', 'ethnicities' and 'discourses'. As the units for analysis for the study are online (either websites or discussion forums), there must be recognition of the multimodal affordances such settings present. Much of the text featuring on the sites is itself multimodal as it makes use of the technical functionality associated with online text. What accompanies the text also uses a range of features particular to online environments (images – still and moving – audio, hyperlinks, 'breadcrumbs', layout). How these varied aspects of online environments operate upon the reader/interlocutor and how they combine to make and enhance meaning is the task of the analyst, recognising the subjective and limited range of interpretations any one analysis may bring to the party. The discourse analysis is thus multimodal as it seeks to account for a holistic interpretation of what is going on while happy at times to focus on identifiably separate strands and delve into particularly interesting nooks and crannies of the data.

Taking bay'ah – pledging allegiance to a spiritual guide in the Islamic tradition: A case study of the sociolinguistic implications of enacting rituals online

In this chapter, I share data gathered from sites that present and facilitate an example of 'online religion', the online enactment of ritual.

Online ritual as a social and religious phenomenon, and as a subdivision of broader inquiry into digital religion, has been researched in a number of different faith contexts. Jacobs (2007) has investigated the posting of online prayers in a Christian Virtual Church and performances of 'cyberpuja' in Hindu virtual temples. Radde-Antweiler has focused on magic rituals in her analysis of Wiccan websites. MacWilliams (2006) has explored controversies surrounding the 'techno-ritualisation' of the Nichiren Buddhist Gohonzon. In this volume, Pandhariphande explores the linguistic implications for online *puja* and *satsang*. These studies have tried to address various questions including issues of space, time, authority, performance and roles.

The principal characteristics of online ritual are its asynchronicity, the emergence of a new 'space' ('sacred cyberspace') for ritual performance, the virtual shrinking of physical distance, its shift to an individual rather than a collective practice, the 'flattening out' of hierarchies and the modification of signs or symbols of authority.

The case study presented in this chapter involves the Islamic ritual of *bay'ah*, or allegiance-pledging, found most commonly among the mystical 'fraternities' across the Islamic world known as *turuq*, often translated into English as 'Sufi Orders'. Muslims claiming membership of such networks often practise a contemplative faith involving a significant amount of supplementary prayer, the recitation of litanies and a range of ascetic and ecstatic practices.

The ritual of *bay'ah* involves an aspirant (or *murid*) pledging allegiance to a spiritual guide or teacher (*shaykh* or *murshid*). This oath commits the aspirant to the teaching and guidance of the guide for as long as necessary, often a lifetime.

The ritual as traditionally (offline) performed involves physical contact (usually taking the hand of the *shaykh*) and the recitation (individually and collectively) of prescribed formulae, some of which are verses from the Qur'an and others which form part of a particular *tariqa*'s textual repertoire. Authority is inherent in the person of the *shaykh* (or his representative) and the sacred nature of the words uttered. Another relevant feature of the offline[1] ritual is that, although it can take place anywhere, participants might travel great distances (often across the globe) to meet with their guide and pledge their allegiance.

This then is traditionally a very personal, physically intimate and significant ritual involving substantial personal investment. Via the internet, what has normally been an essentially oral collective performance involving close physical proximity has now become – in certain contexts – an online individual practice, primarily literacy based, asynchronous and physically separated. The linguistic implications of this transfer are interesting and none more so than the shift that takes place when rituals go online from an oral-based performance to a literacy-based one. However, there are other matters of interest including, in the cases shared in this chapter, the uncertainties inherent in the varied use of transliteration and transcription, and the multilingualism of these online spaces including the emergence of Islamic varieties of English. Where written text is directly transliterated from one script to another there is obviously a strong literacy orientation and less possibility for variation. On the other hand, where aurally experienced text is *transcribed* rather than *transliterated*, orientation is very much an oral one and admits more extensive variation both in style and in accuracy. Not infrequently, there is an admixture of both transliteration and transcript within the same text, examples of which appear below.

Some of this activity can be subsumed or interpreted through 'secondary orality' as defined by Ong (1982), as 'essentially a more deliberate and self-conscious orality, based permanently on the use of writing and print'. Web 2.0 applications and platforms, in particular, have led to a massive explosion in secondary orality-oriented activity, to the extent that some argue that the age of print has been only an interruption, a 'Gutenberg Parenthesis' (Pettit, 2010), to the more coherent progression of language practices which are at root oral and multisensory. Online ritual with its residual orality and its digital placeholder is an example of such a process.

The data

The data in this chapter comes from three different online settings. They consist of a web page each from three websites linked to two of the most well-known global Sufi *tariqas* – the *Shadhili* and the *Naqshbandi*. These networks are transnational (particularly in the modern age) and often adopt English as the lingua franca for their participants.

A *tariqa* (lit. 'path') is the Arabic term usually employed to denote the often large and increasingly transnational organisations based around the teachings of a particular spiritual guide, who tends to be the latest successor in a long line of similar guides stretching all the way back to the time of the Prophet in 7th-century Arabia. These two *tariqas* are named after particularly famous guides from the 13th and 14th centuries (CE), respectively, who developed and consolidated the practices of each particular path.

I am using the Arabic term *tariqa* deliberately to foreground one of the major issues arising from an analysis of these online rituals and their associated websites, namely the discursive tension around conveying technical terms denoting religious practices which are both sourced from, and translated into, a range of different traditions. This discursive tension is highlighted in these online spaces which, by their nature, encourage a wide range of interpretations dictating language choices. The word, *tariqa*, for example, is used on these English language websites (a) in its original Arabic form (transliterated), (b) translated into English by the word 'path' (its literal meaning) and (c) by the etymologically unrelated term 'order' or 'Sufi Order'. The latter two options represent this key tension in the rendering of Islamic terms and concepts into the English language and form part of the emergence of varieties of Islamic English rather than a single standardised variety. This tension partly consists of the use of what one might call 'Eastern/New Age' terms and concepts or language forms more associated with the Judeo-Christian tradition (via Orientalism) – particular the co-sanctified (Fishman, 2006) analogous forms arising from translations of the Bible such as the authorised King James version from the early 17th century (KJV, 1611/2011). Co-sanctification is the term used by Fishman (2006) to denote language varieties which, through particular usage in sacred or religious contexts, have taken on the 'aura' generally associated more with sacred languages such as Qur'anic Arabic, Church Slavic and Biblical Hebrew or Aramaic. Urdu and Persian, for example, in certain religious registers and domains, particularly poetry, have attracted these notions of sanctity for the language they use. Similar patterns of Judeo-Christian English are found in 19th-century translated texts from Hinduism and Buddhism (Yelle, 2013).

Analysis

The procedure followed in this study was, first of all, to identify two online sites where allegiance-pledging was possible and aimed at an English-speaking audience. In the end, I chose to use three such sites but a quick Google search provided another half dozen or so similar ones from different organisations, including Arabic language only sites.

The relevant page was screen grabbed and printed off on an A4 sheet of paper and then stuck onto a larger A3 sheet of paper. An analysis was conducted in the following manner. (1) A tally was made of different languages and varieties used on each page; (2) key words and phrases were identified based on my working hypothesis regarding different varieties of Islamic English; (3) words and phrases were assigned to different repertoires; (4) a close analysis was undertaken on a syntactic and lexical level about the convergence of language and other linguistic resources (e.g. scripts) on the page; and (5) examples were identified of where oracy and literacy practices coexisted in an often awkward combination.

'Unto the One'

The first site (Figure 10.1), Unto the One, is a website which allows for the pledging of allegiance to a teacher in the *Shadhili tariqa* ('Unto the One', n.d). There are 151 words in total on the page, of which only 5 are discrete Arabic words (glossed immediately) and the rest are in English. Some of these are loanwords with an Arabic origin (such as 'Sufi' or 'S/sheikh').

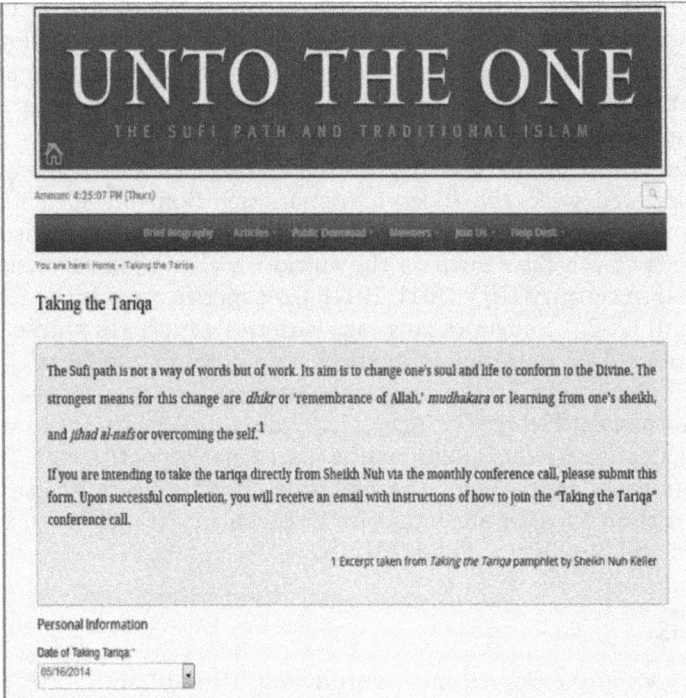

Figure 10.1 From the website of a *Shadhili* 'tariqa' or 'order'

There are a small number of proper nouns, mainly names, also of Arabic origin ('Nuh', 'Amman').

The first noteworthy thing to point out is the name of the page and website, 'Unto the One'. The choice of an obviously biblical variety of English here is a typical example of how emerging varieties of Islamic English have dealt with translating or conveying Arabic terms, or, in this case, alighting on an English phrase with no direct equivalent in Arabic but which indexes a variety of English recognisably religious (or sanctified).

A similar expression is '...conform to the Divine' which, as an example of biblical phraseology, links to the title of the page 'Unto the One' with its connotations of 'No one comes unto the Father...' or even the more secular Shakespearian 'Once more unto the breach'. In Fishman's (2006) schema, this is a clear example of leakage from one existing religious variety to newly emerging one(s) operating across faith boundaries. The co-sanctification of the language of the King James Bible and the emerging literary standard in England at the same historical moment combine to give these English phrases a quasi-religious register.

In respect of typical lexical items characteristic of such sites, the Arabic word *tariqa*, which is central to this particular web page, is either left as such or translated as 'path'. There is no occurrence of the Judeo-Christian 'order' (see later). It is, however, attached to the adjective *'Sufi'*, perhaps to signal its particular usage in this context. For an Arabic speaker, *tariqa* can be understood without the corresponding adjective, though in more formal Arabic contexts, it might also be attached.

Of interest here is the use of the English verb 'take' in the phrase 'taking the tariqa' which is used here to denote the pledging of the oath of allegiance to a guide, or *shaykh*. This appears three times, once in the body text, once attached to the 'conference call' instruction and also as the name of a pamphlet. This is possibly a direct translation from the Arabic of *akhatha al-tariqa* and signals a much closer relationship to the source language from the designers of the website than from some others.

By contrast, a closer analysis of the language on this page reveals very idiomatic and accurate Standard and, in places, literary English, revealing perhaps the native-speaker origins of the teacher in question or the website designers. There is a minimum of Arabic terminology which when used is glossed immediately (*dhikr, mudhakara, jihad al-nafs*) and therefore makes no assumptions about visitors' prior knowledge. The juxtaposition of this text with web conventions such as 'breadcrumbs', automated form-filling and various headers and footers make for a hybrid text which brings this age-old ritual into the world of electronic communication.

The move from orality to literacy is here signalled by a traditional ritual activity (taking *bay'ah*) involving a formal registration (online form-filling) which generates an email and then a monthly conference call – which, as its name suggests, is not a one-to-one event, though an element of immediacy and oracy is retained here by the telephone call. The hybrid form of this event consists then of elements of online communication combined with an effort to retain the oral nature of the ritual through the conference call with the teacher at a certain time during the week. There is a blend here of synchronicity and asynchronicity.

A second major factor here, represented by the title of the page as well as elsewhere, is, of course, the continuing sociology of the language and religion concept of co-sanctification.

Naqshbandi-Haqqani Tariqa

The second site (Figure 10.2) is from the *Naqshbandi-Haqqani Tariqa* ('Initiation', n.d.). The allegiance-pledging pages are two in number, linked and lengthier. There is significantly more bilingualism and biliteracy on these pages than with Unto the One. Key words here are 'Initiation', *'Tariqa'* and *'Bay'ah'*. Here, the intention of the page seems to signal the concept of 'initiation'. The English translation of *bay'ah* is not strictly speaking 'initiation' because *bay'ah* means swearing an oath of allegiance. The use of this word, however, is an example of the maintenance of Judeo-Christian

Home   The Tariqa ⌄   Teachings ⌄   Golden Chain ⌄   Practices

Initiation

Taking Bay'a - Initiation

For those people who cannot reach one of our
authorized representatives, the Shaykh has granted
permission to take initiation by reciting the Baya'
text along with this recording.

أشهد ان لا اله الا الله , وأشهد أن محمدا رسول الله

أشهد ان لا اله الا الله , وأشهد أن محمدا عبده ورسوله (2 x)

بسم الله الرحمن الرحيم

إنَّ الَّذِينَ يُبَايِعُونَكَ إِنَّمَا يُبَايِعُونَ اللَّهَ يَدُ اللَّهِ فَوْقَ أَيْدِيهِمْ فَمَن نَّكَثَ فَإِنَّمَا يَنكُثُ عَلَى نَفْسِهِ

وَمَنْ أَوْفَىٰ بِمَا عَاهَدَ عَلَيْهُ اللَّهَ فَسَيُؤْتِيهِ أَجْرًا عَظِيمًا

ashadu an lā ilāha ill 'Llāh wa ashadu anna Muhammadan rasoolullāh – I bear
witness that there is no god except Allah and I bear witness that Muhammad is the
Prophet of Allah.

ashadu an lā ilāha ill 'Llāh wa ashadu anna Muhammadan 'abduhu wa rasūluh (2x) –
I bear witness that there is no god except Allah and I bear witness that Muhammad
is His servant and Prophet.

bismillāhir-rohmān ir-rahīm – In the name of Allah the Compassionate, the Merciful.

*Inn'alladhīnā yuba'ūnaka innamā yuba'ūn-Allāh, yadullāhi fawqa aydihim, faman
nakatha fa-innamā yankuthu 'alā nafsih wa man awfā bimā 'ahad 'alayhullaha
fa-sayu'tīhi qiran 'azhīma*

Lo! those who swear allegiance unto thee (Muhammad), swear allegiance only unto
Allah. The Hand of Allah is above their hands. So whosoever breaketh his oath,
breaketh it only to his soul's hurt; while whosoever keepeth his covenant with Allah,
on him will He bestow immense reward. [Surat al-Fath, 48:10]

Figure 10.2 From the website of a *Naqshbandi* 'tariqa' or 'order'

(via Orientalism) terms where 'initiation into an order' reflects either a
monastic or esoteric Western tradition of magical and secret 'orders'. A
quick Google search using the phrase 'initiation' and 'order' gives us these
as the first 10 hits (Figure 10.3). Only one of these hits (the sixth one) refers
to a *tariqa* in the sense used here. The rest are mainly a collection of sources
explaining initiation into esoteric organisations.

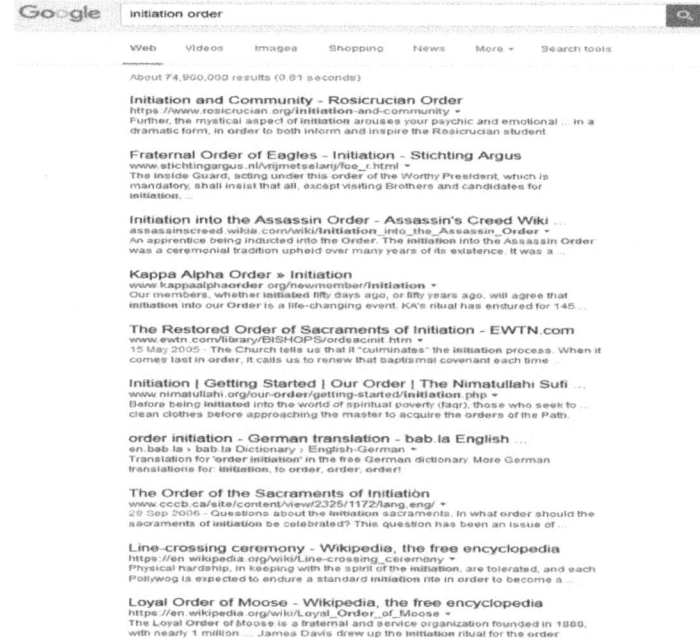

Figure 10.3 List of top 10 sites following Google search of 'initiation' and 'order'

Seminal orientalist texts such as Lane's (2010) *Manners and Customs of the Modern Egyptians* (published in 1836) used such terms when describing the activities of the *'daraweesh'* ('dervishes', followers of a Sufi teacher).

Unlike Unto the One, swearing allegiance here is a ritual that takes place exclusively online, asynchronously, moving performance from an oral performance into an exclusively online literacy performance involving reading/decoding and recitation.

There is, like Unto the One, an accompanying form-filling element to the ritual (again more literacy). There is a collage of images including one of the teacher and guide and images of 'real', offline, *bay'ah* taking place (represented by the hands in the images). This is reproduced virtually by the potential aspirant being asked to 'click on the hands' (another aspect of web literacy) in the photo in order to continue to the next stage of the online ritual performance.

On this page and the next, we find the expressions 'takes *bay'ah*', 'to take initiation' and 'to make *bay'ah*'. These variants reveal a certain lexical inconsistency and instability as this oral ritual is conveyed in the English language. The two websites so far have dealt with this in different ways – Unto the One choosing to calque the Arabic with 'taking the *tariqa*' and the *Naqshbandi-Haqqani* site both sidestepping Arabic with the use of the

Orientalist 'initiation' and using the loanword 'making' or 'taking *bay'ah'*. All this reminds us of the challenges posed by translating technical terms from one faith tradition to another and that using existing terms in the target language does not necessarily convey the meaning of the source term and can indeed index different concepts altogether. The use of the word 'initiation' is an example of this (see Pandharipande in this volume for examples in the Hindu tradition; see Ostler [2016] for a book-length discussion of how translation can potentially lead to a mismatch of religious concepts).

The main page contains a downloadable audio file and the text of what is to be recited as part of the online ritual swearing of allegiance. The texts are in Arabic and English with the Arabic appearing first in the original Arabic script followed by a transliterated/transcribed hybrid, and then a translation into English. The hybrid nature of the transliteration/transcription is reliant on a number of factors. There are considerable challenges to any transliteration of Classical Arabic text, as beyond the accurate representation of letters, there are important pronunciation conventions which are not always denoted or often not appreciated by readers. Different systems of transliteration, therefore, have been developed which reflect this. Another factor is the orality/literacy distinction, with numerous non-Arabic-speaking Muslims familiar with many other Arabic terms and formulae only in an oral way and therefore are used to more *ad hoc* methods of transcription (rather than transliteration) of the oral sounds. We see a mixture of all these in the text.

The English translation used here (Pickthall, 1930) adopts the lexical and syntactical style of the King James translation of the Bible, emphasising the co-sanctification of language, with the use of 'bear witness', 'Lo', 'thee', 'unto', whosoever', 'breaketh', 'on him will He bestow' indexing an archaic variety with archaic pronouns, syntax and lexical choices. This is a variety of Islamic literary English that is strongly rooted in an Orientalist tradition which here denotes the mediation of one religion (Islam) through the language associated with a different religion (Christianity in its broadly Western variety). Later, we will see how the same or similar Islamic concepts are denoted in the English language via more Eastern traditions – again though via Orientalism.

This page is a placeholder for a number of varieties of written English – the very contemporary register of web conventions such as the clickable link 'please fill in the Detailed Contact Form or email us', the archaic and co-sanctified English translation of the Qur'anic verse and the hybrid transliteration/transcription and diacritically marked Arabic.

There is regular use of the English word 'Order' and links to the comments made above about 'initiation'. Here, the web page writer/designer has elected to adopt the stock term 'Order' or 'Sufi Order' to denote the Arabic 'tariqa'. This complements the co-sanctified terms employed for the translation of the Qur'anic verse. The long translated text towards the end is the dedication text, a translingual text using names and titles from Arabic and Persian (*shah, Khawaja*) and Arabicised names. There is also inconsistency in the use of 'Allah' and 'God'.

'Taking *Bayyath*'

The third site (Figure 10.4) illustrates different linguistic influences on the English language features of these pages. 'Taking *Bayyath*' is quite different from the previous two even though it is linked to the same teacher as the second site. The lexical choices in evidence here often reflect a more 'Eastern' or 'New Age' register with words and phrases such as 'an alternative method', 'connect up', 'open up spiritually', 'to progress spiritually', 'connection', 'to attain to spiritual fulfilment', 'spiritual openings' and 'spiritual advancement' as typical. They also betray the South Asian provenance of the site managers with certain transcription conventions representing South Asian language users' rendering of Arabic

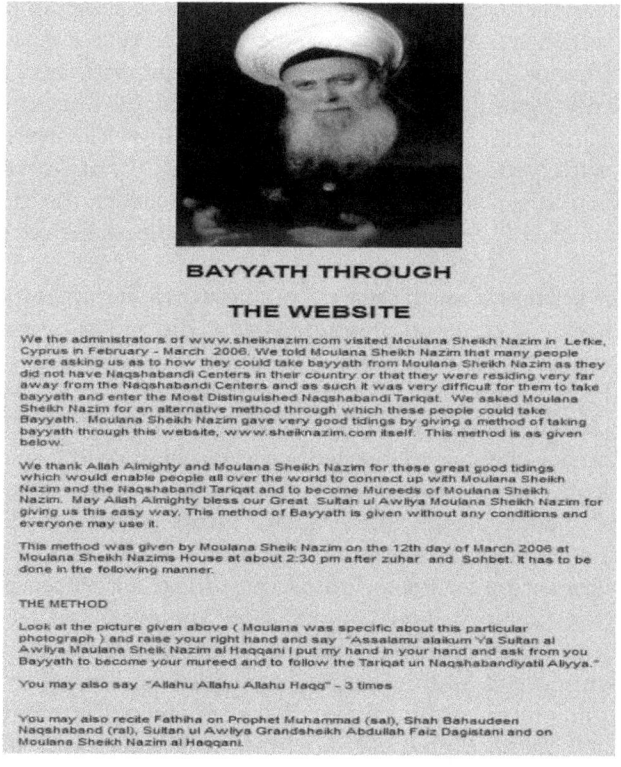

Figure 10.4 From another website of a *Naqshbandi* 'tariqa' or 'order'

terms. For example, *bay'ah* here is transcribed/transliterated (it's not clear which) as *bayyath*.

Unlike in the previous site, most of the formulas to be recited in the online ritual are in English rather than in Arabic and use key lexical items associated with *tariqas*. Arabic loanwords such as *'mureed'* (aspirant) and *'mowlana'* (our master) are used as is the name of the tariqa itself as *'Tariqat un Naqshabandiyatil Aliyya'* (attempting to transliterate pronunciation conventions). But there is a minimalism here akin to the first site Unto the One. The ritual is performed online, asynchronously, using a physical gesture (raising of the right hand in front of the photo on the screen) and the normal Islamic greeting of *'Assalamu alaikum'* (Peace be upon you).

However, within this generally Eastern – or New Age – oriented discourse there are also elements of the more Judeo-Christian discourse evident in the first two sites. Phrases such as 'May Allah Almighty bless...', 'We thank Allah Almighty...' and '... grant good tidings...' represent a more Western and Judeo-Christian-inflected register. Interestingly, in the area of Islamic spirituality, or Sufism, such discursive elements often contrast with discourses associated with either Eastern (Hindu, Buddhist) practices or New Age movements, both spiritual and secular. We see examples of both in 'spiritual fulfilment' and 'alternative method' and such a tension perhaps represents different cultural emphases and local linguistic resources rather than necessarily signalling different religious practices. This can take place sometimes in the same text as in 'spiritual openings' (Eastern/New Age) contrasting with 'good tidings' (Judeo-Christian). Although both this site and the previous one claim to represent the same teacher, their cultural and linguistic emphases as well as their methods of pledging allegiance are quite different.

Again, we see here a much greater emphasis on a literate response rather than an oral one. There is no downloadable audio file to provide an aural experience and little to recite aloud. There is, for example, omission of most official Arabic formulae including the relevant Qur'anic verses which accompany pledging allegiance to a teacher.

Even a partial analysis of the language in these online spaces reveals a patchwork of inconsistency and provisional admixture of different scripts, varying conventions and language forms which betray a range of sources feeding each page. These include Judeo-Christian language forms, Orientalist terminology ('order', 'initiation'), the orality/literacy hybridity present in transliteration and transcription and web-based conventions such as downloadable audio files, 'clicking', email links and other interactive functions common to online spaces in general. Attempts are made to simulate offline practice with images of the teacher and aspirants, clicking on images to symbolise physical proximity and sound files to make the experience of online ritual a multisensory one. However, all the identifiable characteristics of online ritual are present – asynchronicity, literacy over oracy, the virtual shrinking of distance, individual over collective experience and the flattening out of hierarchies, e.g. there is less possibility of being 'vetted' to be an acceptable aspirant.

These are transnational platforms which use the English language as a lingua franca and the different ways in which English is used index the varied sources these platforms draw upon and their cultural and religious emphases. For example, an apparent tension between Judeo-Christian and Eastern/New Age discourses characterises all three sites. The provisional and inconsistent nature of these language practices relate to a number of factors. Obviously, online language practices appear to encourage a bricolage approach to genres, conventions and language forms. This hybridity manifests itself in a number of different ways. The uncertainty between transliterated and transcribed forms reflects the transition of hitherto oral practices into literacy-oriented contexts where potential devotees call upon different language skills and knowledge in order to participate. On religious sites where sacred languages such as Classical Arabic are present, or languages more traditionally linked to Islam than to English, such as Urdu, issues around the use of original script, transliteration, transcription and translation emerge. For example, if Arabic is recited with a particularly South Asian inflection then this may carry over into the transcription. If certain prayers or oral formulas are learned orally through memorisation, rather than through the written word, then transcription can clash with transliteration, which may be more faithful to the original. Both may appear in the same text as in the *Naqshbandi-Haqqani Tariqa* example above. By the same token, some website writers may opt for more Judeo-Christian English language forms in the belief there is a co-sanctification dimension to such language. Others may deliberately choose the English language forms associated with other Eastern and New Age religions to denote particular devotional acts, as in the Al Bayyath example above. They may even include both in the same text.

The hybridity on these religious sites designed to facilitate pledging allegiance to particular teachers is therefore multifaceted and complex. Mode (orality/literacy), religious orientation (Judeo-Christian/Eastern/New Age), language (varieties of English, Classical Arabic, localised varieties of recited Arabic and scripts) and presentation (transcription/transliteration/translation) all interplay resulting in dynamic and evolving genres. To what extent this hybridity and its constituent parts impacts on religious concepts and practice is difficult to judge. There is a much clearer impact on the latter as the move to online spaces inevitably leads to instances of asynchronous and individualised practice. As Ostler (2016) concludes, in a far wider ranging study than this chapter, the move, in his case, of a faith from one linguistic context to another can create spaces for subtle and nuanced reconceptualisations. The same could be said for the move of a faith to an online context. However, to what extent these reconceptualisations are significant in the lives of practitioners is less clear.

Conclusion

A guiding hypothesis for this study has been the observation that when an offline ritual performance moves online, an accompanying shift from oral performance to literacy performance accompanies it. The data shared in this chapter capture some of that transformation. They also capture some of the hybrid and converging modes of communication evoked by online practices. Whether or not the latter develop further to significantly impact the development of both online and offline language practices and forms, and our theorisation of them, remains hard to ascertain. Needless to say, the examples included here have their parallels across other contexts, faith and non-faith oriented, as well as across other online platforms and, no doubt, future platforms yet to be developed. The intimate relationship between faith and language will undoubtedly be affected by the emergence of online spaces in the Information Age (Castells,1996/1998). Whether or not we ever reach Strate's (1996: 373) state of 'new consciousness' that 'may emerge through a synthesis between our physical selves and the dream selves we generate in cybertime', there is the tantalising prospect that potential new understandings of language may transform experiences and conceptualisations of religious practice.

Note

(1)   The fact that we now use 'offline' as one part of a contrastive pair with 'online' to denote what was once a single, physical world practice, is telling.

References

(KJV), H.B. (2011 (1611)) *Holy Bible: King James Version (KJV)*. London: Collins.

Baker, J.N. (1995) The presence of the name: Reading scripture in an Indonesian village. In J. Boyarin (ed.) *The Ethnography of Reading* (pp. 98–138). Berkeley, CA: University of California Press.

Baron, N. (2003) Why email looks like speech: Proofreading, pedagogy, and public face. In J. Aitchison and D. Lewis (eds) *New Media Language* (pp. 102–113). London: Routledge.

Baron, N. (2007) Text messaging and IM: Linguistic comparisons of American college data. *Journal of Language and Social Psychology* 26, 291–298.

Baron, N. (2008) *Always On: Language in an Online and Mobile World*. New York: Oxford University Press.

Barton, D. and Hamilton, M. (1998) *Local Literacies: Reading and Writing in One Community*. London: Routledge.

Bauman, R. (1974) Verbal art as performance. *American Anthropologist* (77), 290–311.

Bauman, R. and Briggs, C. (1990) Poetics and performance as critical perspectives on language and social life. *Annual Review of Anthropology* 19, 59–88.

Bledsoe, C.H. and Robey, K.M. (1993) Arabic literacy and secrecy among the Mende of Sierra Leone. In B. Street (ed.) *Cross-Cultural Approaches to Literacy* (pp. 110–134). Cambridge: Cambridge University Press.

Bock, Z. (2015) 'Why can't race just be a normal thing?': Entangled discourses in the narratives of young South Africans. *Working Papers in Urban Language & Literacies*.

Bolter, J.D. (1991) *Writing Space: The Computer, Hypertext, and the History of Writing.* Hillsdale, NJ: Lawrence Erlbaum and Associates.

Boyarin, J. (1995) Placing reading: Ancient Israel and medieval Europe. In J. Boyarin (ed.) *The Ethnography of Reading* (pp. 98–138). Berkeley, CA: University of California Press.

Butler, J. (1997) *Excitable Speech: A Politics of the Performative.* New York: Routledge.

Campbell, H. (2007) Who's got the power? Religious authority and the internet. *Journal of Computer-Mediated Communication* 12 (3), 1043–1062.

Campbell, H. (2010) Religious authority and the blogosphere. *Journal of Computer-Mediated Communication* 15 (2), 251–276.

Campbell, H. (2012) Understanding the relationship between religion online and offline in a networked society. *Journal of the American Academy of Religion* 80 (1), 64–93.

Campbell, H. (2013) *Digital Religion: Understanding Religious Practice in New Media Worlds.* Abingdon: Routledge.

Casey, C. (2006) Virtual ritual, real faith: The revirtualization of religious ritual in cyberspace. *Heidelberg Journal of Religions on the Internet* 2 (1).

Castells, M. (1996/1998) *The Information Age: Economy, Society and Culture* (3 vols). Oxford: Wiley-Blackwell.

Conquergood, D. (2013) *Cultural Struggles: Performance, Ethnography, Praxis.* Ann Arbor, MI: University of Michigan Press.

Crystal, D. (2006) *Language and the Internet.* Cambridge: Cambridge University Press.

Crystal, D. (2008) *Txtng: The Gr8 Db8.* Oxford: Oxford University Press.

Danet, B. (2001) 'Feeling spiffy': The changing language of public email. In D. Slater (ed.) *Cyberpl@y: Communicating Online* (pp. 51–99). Oxford: Berg Publishers.

Delaney, P. and Landow, G. (1991) *Hypermedia and Literary Studies.* Cambridge, MA: MIT Press.

Digges, D. and Rappaport, J. (1995) Literacy, orality and ritual practice in highland Colombia. In J. Boyarin (ed.) *The Ethnography of Reading* (pp. 139–155). Berkeley, CA: University of California Press.

Eigenbrod, R. (1995) The oral in the written: A literature between two cultures. *The Canadian Journal of Native Studies* 15 (1), 89–102.

Eliade, M. (1965) L'initiation et le monde moderne. In C.J. Bleeker (ed.) *Initiation* (pp. 1–14). Leiden: E.J. Brill.

Febvre, L. and Martin, H.J. (1976) *The Coming of the Book: The Impact of Printing 1450–1800.* London: Verso.

Fishman, J.A. (2006). A decalogue of basic theoretical perspectives for a sociology of language and religion. In T. Omoniyi and J.A. Fishman (eds) *Explorations in the Sociology of Language and Religion.* (pp. 13–25). Amsterdam: John Benjamins.

Gee, J.P. (1990) *Social Linguistics and Literacies: Ideology in Discourses.* London: Falmer Press.

Gillen, J. (2000) Recontextualization: The shaping of telephone discourse in play by three- and four-year-olds. *Language and Education* 14 (4), 250–265.

Goody, J. (1968) *Literacy in Traditional Societies.* Cambridge: Cambridge University Press.

Goody, J. (1977) *The Domestication of the Savage Mind.* Cambridge: Cambridge University Press.

Graff, H. (1979) *The Literacy Myth: Literacy and Social Structure in the 19th Century City.* New York: Academic Press.

Havelock, E.A. (1976) *Origins of Western Literacy.* Toronto: Ontario Institute for Studies in Education.

Helland, C. (2007) Diaspora on the electronic frontier: Developing virtual connections with sacred homelands. *Journal of Computer-Mediated Communication* 12 (3), 956–976.

Hopper, R. (1992) *Telephone Conversation.* Bloomington, IN: Indiana University Press.

Initiation (n.d.). Initiation. The Naqshbandiyya Nazimiyya Sufi Order of America. See www.naqshbandi.org/the-tariqa/initiation/ (accessed December 2016).

Jacobs, S. (2007) Virtually sacred: The performance of asynchronous cyber-rituals in online spaces. *Journal of Computer-Mediated Communication* 12 (3), 1103–1121.

Joyce, M. (2002) No one tells you this: Secondary orality and hypertextuality. *Oral Tradition* 17 (2), 325–345.

Lane, E.W. (2010 [1836]) *Manners and Customs of the Modern Egyptians.* New York: Cosimo.

Lee, J.L. (2003) Charting the codes of cyberspace: A rhetoric of electronic mail. In In L. Strate, R. Jacobson and S. Gibson (eds) *Communication and Cyberspace* (pp. 307–388). Cresskill, NJ: Hampton Press Inc.

Lord, A.B. (1960) *The Singer of Tales.* Cambridge, MA: Harvard University Press.

MacWilliams, M. (2006) Techno-ritualization: The Gohozon controversy on the internet. *Heidelberg Journal of Religions on the Internet* 2 (1).

McLuhan, M. (1962) *The Gutenberg Galaxy: The Making of Typographic Man.* Toronto: University of Toronto Press.

Noakes, S. (1995) Gracious words: Luke's Jesus and the reading of sacred poetry at the beginning of the Christian era. In J. Boyarin (ed.) *The Ethnography of Reading* (pp. 38–57). Berkeley, CA: University of California Press.

O'Leary, S.D. (1996) Cyberspace as sacred space: Communicating religion on computer networks. *Journal of the American Academy of Religion* 64 (4), 781–808.

Ong, W.J. (1982) *Orality and Literacy.* London: Routledge.

Ostler, N. (2016) *Passwords to Paradise: How Languages Have Re-Invented World Religions.* London: Bloomsbury.

Parry, A. (1971) *The Making of Homeric Verse: The Collected Papers of Milman Parry.* Oxford: Clarendon Press.

Pettitt, T. (2007) Before the Gutenberg parenthesis: Elizabethan American compatibilities. Media in Transition 5: Creativity, Ownership and Collaboration in the Digital Age. Communications Forum, Massachusetts Institute of Technology, Cambridge, MA.

Pickthall, M.W. (1996) *The Meaning of the Glorious Koran: An Explanatory Translation.* Amana Publications. (Originally published 1930.)

Probst, P. (1993) The letter and the spirit: Literacy and religious authority in the history of the Aladura movement in western Nigeria. In B. Street (ed.) *Cross-Cultural Approaches to Literacy* (pp. 198–219). Cambridge: Cambridge University Press.

Purcell-Gates, V. (1998) Growing successful readers: Homes, communities, and schools. In F. Lehr and J. Osborn (eds) *Literacy for All: Issues for Teaching and Learning* (pp. 51–72). New York: Guilford Publications.

Radde-Antweiler, K. (2006) Rituals online: Transferring and designing rituals. *Heidelberg Journal of Religions on the Internet* 2 (1).

Reder, S. and Wikelund, K.R. (1993) Literacy development and ethnicity: An Alaskan example. In B. Street (ed.) *Cross-Cultural Approaches to Literacy* (pp. 176–197). Cambridge: Cambridge University Press.

Schechner, R. (1988) *Performance Theory.* New York: Routledge.

Schechner, R. (2002) *Performance Studies: An Introduction.* London: Routledge.

Scheifinger, H. (2013) Hindu worship online and offline. In H. Campbell (ed.) *Digital Religion: Understanding Religious Practice In New Media Worlds* (pp. 121–127). Abingdon: Routledge.

Scribner, S. and Cole, M. (1981) *The Psychology of Literacy.* Cambridge/London: Harvard University Press.

Standage, R. (2014) *The Victorian Internet: The Remarkable Story of the Telegraph and the Nineteenth Century's On-Line Pioneers.* New York: Bloomsbury.

Strate, L. (1996) Cybertime. In L. Strate, R. Jacobsen and S. Gibson (eds) *Communication and Cyberspace: Social Interaction in an Electronic Environment.* (pp. 351–378). Cresskill, NJ: Hampton Press.

Strate, L., Jacobsen, R. and Gibson, S. (1996) *Communication and Cyberspace: Social Interaction in an Electronic Environment.* Cresskill, NJ: Hampton Press.

Street, B. (1984) *Literacy in Theory and Practice.* Cambridge: Cambridge University Press.

Taking Bayyath (n.d.) Taking Bayyath. The Sheikh Nazim website. See www. sheiknazim2.com/Bayyath.html (accessed December 2015).

Taking the Tariqa (n.d.) Taking the Tariqa. Unto the One: The Sufi path and traditional Islam. See www.untotheone.com/taking-the-tariqa/ (accessed December 2015).

The Holy See (n.d.) The Holy See. See www.w2.vatican.va/content/vatican/en.html (accessed December 2015).

Turner, V. (1969) *The Ritual Process.* Chicago, IL: Aldine.

Weber, M. (2002) *The Protestant Ethic and the Spirit of Capitalism* (G. Wells, trans.). London: Penguin.

Yelle, R. (2013) *The Language of Disenchantment: Protestant Literalism and Colonial Discourse in British India.* Oxford: Oxford University Press.

Part 5

Afterword

11 Afterword

Bernard Spolsky

In a book without an editor's introduction, the Afterword is an opportunity to summarize and comment on the contributions. Here rather, I am allowed more freedom to add, and choose therefore to start with my personal experience, and then, after discussing in some detail the way that digital media have encouraged exploration and exploitation of Jewish religious observances, to ask briefly whether this approach might also fit the other chapters in this book.

First, the personal. For many years, whenever I travelled I carried with me a Hebrew prayer book so that I could take part in the formal observances that take place when 10 or more observant Jewish males find themselves together at an appropriate time. Now, like many others, I rely on the digital program on my iPhone. Thus, the availability of a digital application with a complete daily prayer book has made it easier to remain observant. Another personal example: for the 30 years that I lived in the Old City of Jerusalem, I would attend a daily Talmud class conducted by a local rabbi; there were normally a dozen other participants ranging in age and experience. The teacher regularly recorded his lesson, which was then stored online and made available to a group of his followers in South Africa. Since I have moved away from the neighborhood, I can access the full collection of Talmud lessons, choosing to listen to the assigned page for the day on my computer at any time that is convenient to me. In addition to sessions on the Talmud, the rabbi records his regular lessons on the weekly biblical portion and on seasonal religious observances. While most of this material is only available on audio, from time to time the lesson is also videotaped, so that one can have the feeling of participation.[1]

Essentially, what these examples show, confirming the many fascinating studies in this volume, is that the availability of digital media has permitted freer and wider access to religious observance and study, breaking down the limitations of time (I can choose when to look at a lesson) and space (I can access the material from my home or from anywhere that I have internet access). I should also note the limitations: an observant Jew follows the prohibition against using electricity (or electronics) on the Sabbath and other holy days. Thus, when I go to the synagogue on the Sabbath, I use a traditional prayer book, and if I wish to study on the Sabbath, I use a book rather than my computer.

But this development also signals one of the important characteristics of digital media. Whereas television and radio (with the exception of local FM radio[2]) tend to be mass media, dependent on large audiences to justify their existence, the internet now encourages the development of quite small networks. This phenomenon, referred to as the long tail, was first identified by Anderson (2004, 2006), and means that it can be used with small (the two dozen learners in Jerusalem and Johannesburg) as well as large groups. As the chapters in this volume illustrate, digital media have been exploited by various religious groups and leaders as their potential has become clearer.

In a further development, access to online classes led to the establishment in 2007 of *Webyeshiva*, a virtual yeshiva offered online, which caters for several thousand participants offering a full range of Torah classes in Hebrew, English, Russian and Spanish. Given the importance of teaching and learning to Judaism after the destruction of the Temple,[3] digital media offer an obvious method of carrying on this aspect.

In his foundational article on the sociology of religion and language, Fishman (2006: 18, italics in original) did not deal with digital media specifically, although item (vi) prepares for it: *'All sources of sociocultural change are also sources of change in the sociolinguistic repertoire vis-à-vis religion, including religious change per se'*. In my earlier examples, the availability of online resources has permitted the spread of Jewish religious teaching beyond the local community and synagogue, providing a vast range of material previously unavailable to the homebound. In the Jewish case, it does not as far as I know include formal worship,[4] but covers the whole range of learning from formal Biblical and Talmud classes to sites like *Mi Yodeya*, one of the Stack Exchange QA sites, intended 'for those who base their lives on Jewish law and tradition and anyone interested in learning more'.[5] Of course, the former vary in the quality and reputation of the institutions or rabbis who write the lessons, and the QA sites do not claim to provide the locally relevant answers to religious questions that require a rabbinic authority to take into account local conditions.[6] The availability of electronic databases has made it possible to make not just sacred texts but also thousands of questions and rabbinic answers available to the wider community, as in the Bar-Ilan Responsa project (www.biu.ac.il/JH/Responsa/) which makes available a huge collection of published rabbinical writings.

For Judaism, then, digital media have had a major effect in providing wider access to religious learning and knowledge beyond the limitations of face-to-face encounters and the printed material that developed early. And as the studies in this volume have shown, there have been similarly important developments for other religions. The basic story is exploitation of available digital resources for those facets of religious observance that can be performed in this way. The simplest is the passing of texts and lessons, formerly limited to group and individual real-time and place gatherings, something that built up the social relevance and contribution of the religious meeting place (church, mosque, synagogue), and that added to a religious event a possibility of interaction between participants and leaders, thus building new virtual religious communities.

The chapters in this volume report cases of digital media providing support for other aspects of religious identity and observance. Thor Sawin, for instance, deals with the development of virtual local parishes, discussion lists that cater to ethnolinguistic groups (such as Russophone Central Asian Koreans in South Korea, academics and 'Side B' LGBT Christians). These are clearly 'long tail' examples, allowing even greater focusing and specializing than say the provision of different language versions of sacred services at different time.[7]

Another case of a combination of focused material allowing also for the spread of locality is provided by Ana Souza, who reports on Kardecist sites which translate Brazilian lessons of Christian groups believing in reincarnation and spiritualism into the host language for migrants in several countries. Here, the limitation is to teaching texts.

In a wider use, Oladipo Salami discusses the attitude of Yoruba followers of Ifa to the use of online and telephone resources for evangelization, spiritual counseling and live-streaming of church services, sermons and prophetic messages. Here, many of the functions of real religious observance are translated into virtual digital media.

In a related study of a more limited genre, Iyabode Deborah Akande describes how the internet has helped spread the oral literature that incorporates Yoruba taboos, a key element in their tradition and culture.

Televangelism, the use of television by those spreading a religious message, showed up on television after its spread through radio. Shaimaa El Naggar explores an expansion of the spread of Muslim televangelists in the UK and the US using YouTube and social media to reach a wider population.

Rajeshwari V. Pandharipande tackles the more difficult issue faced by Hindus in presenting and accepting two Hindu rituals: *Satsang* (literally, connecting through religious discourses with the Divine/Gurus who have the knowledge and experience of the Divine) and *Puja*, the ritual worship of Hindu deities. Here, the possible limits of the digital media form in allowing for virtual public worship are explored.

In another chapter, Andrey Rosowsky studies the way in which online ritual (in this case the Islamic ritual of *bay'ah*, or allegiance-pledging, found most commonly among the mystical 'fraternities' across the Islamic world known as *turuq*, often translated into English as 'Sufi Orders') crosses the traditional boundaries not just between orality and literacy but also between virtual and real performance.

Finally, the chapter by Tope Omoniyi (co-founder with Fishman of this series of studies of the relations between religion and language) describes examples that illustrate the full range of digital media exploited by various Christian groups. He starts with the famous US television evangelist, Billy Graham, some of whose broadcasts have been preserved and made available on YouTube. This leads to noting the wider use of digital resources for some prayers – the BBC's 'Songs of Praise' and Channel 4's 'Call to Prayer' during Ramadan.[8] Another example he cites is the English and translated programs of the Christian Believers Love World Ministry. There are other such worldwide ministries, including the offering of a Global Communion Service. Thus, digital media allow the virtual presentation of various aspects of religious observance.

One chapter does not fit this pattern, but adds another interesting dimension. Tatjana Soldat-Jaffe discusses Yiddish Wikipedia, used by Yiddish secularists to reverse the loss of the language and build and defend a virtual web community that will continue the struggle of Yiddish as a *mameloshen* (native vernacular) against the traditional *loshn koydesh* (sacred language) Hebrew (or really Hebrew-Aramaic) of Jewish sacred and religious texts. This is a continuation of the struggle at the beginning of the 20th century that reached its apogee in the Tshernovits conference in 1907 (Fishman, 1993) that proclaimed Yiddish as 'a' (and not 'the') language of the Jewish people, and that continued in Ottoman and Mandatory Palestine for many years. One justification for including this topic (it ignores the continued life of Yiddish as the isolating language of some Hasidic sects who practice strict religious observance and use Yiddish as a vernacular and teaching language while keeping Hebrew for liturgical and literary purposes [Katz, 2004]) must be the fact that it was sanctified, as it were, by being the vernacular and literary language of millions of East European Jews murdered by the Nazis and their anti-Semitic supporters in the middle of the 20th century (Fishman, 2002). The argument reappeared, Tatjana Soldat-Jaffe asserts, in the proposal to close the Yiddish Wikipedia site and in the close connection between religious and ethnic identity.

Clearly, just as the study of language and religion which Joshua Fishman pioneered is still in its beginning stages, this volume can offer no more than a set of preliminary explorations of the way that online resources are being exploited in a number of religious contexts. As Rosowsky expressed so clearly in his introduction, this is a challenging topic, to connect faith, language and technology, each well-studied phenomena but now united for the first time, and made even more difficult by its dynamic and developing state. There have been many studies of the enormous effect of writing and later of printing on the development of religion. As Fishman wisely predicted, the addition of a new technology promises to have important effects on religion and language and their relationship. This first collection may be, as its editor claims, a modest contribution, but it is a significant one to which continuing research and publication will add important ways of exploring the connection between religion and language.

Notes

(1) An earlier source of Talmud lessons, following the schedule of the Daf Yomi, a seven and a half year cycle of learning a page of Talmud a day, which began in 1923, was earlier offered over the telephone in a number of countries and languages making it possible for travelers to keep up when away from their home classes.

(2) Church and missionary groups in South America used local FM radio permitting language diversity.

(3) For discussion of the educational and economic effect of this major change for Jews, see Botticini and Eckstein (2012).

(4) One exception is the reading of the Megilla on Purim, a semi-holiday when use of electricity is permitted, on public television programs in Israel.

(5) The *Mi Yodeya* site is monitored by participants who insist on the citing of rabbinic sources for any suggested rulings. There are other sites that do not assure the reader of rabbinic authority.

(6) Traditionally, the answer to whether the accidental mixing of milk and meat made the result no longer permissible was often dependent on the economic status of the questioner: could they afford to replace it? Or in a case when there was a power failure in the Old City of Jerusalem that required Arab electricians to restart individual space heaters on the Sabbath, the rabbinic expert's first question was, are there babies and young children in the apartments? If so, it is permissible.

(7) The Protestant church established in Jerusalem in the 19th century with the purpose of converting Jews had to deal with a dispute over the most popular times for its services in English, Yiddish and Arabic.

(8) Israeli public radio opens in the morning and closes at night with a brief Hebrew prayer or biblical reading.

References

Anderson, C. (2004) The long tail. *Wired Magazine* 12 (10), 170–177.

Anderson, C. (2006) *The Long Tail: Why the Future of Business is Selling Less of More.* New York: Hyperion.

Botticini, M. and Eckstein, Z. (2012) *The Chosen Few: How Education Shaped Jewish History.* Princeton, NJ: Princeton University Press.

Fishman, J.A. (1993) The Tschernovits congress revisited: The first world congress for Yiddish revisited, 85 years later. In J.A. Fishman (ed.) *The Earliest Stage of Language Planning: The 'First Congress' Phenomenon* (pp. 321–332). Berlin: Mouton de Gruyter.

Fishman, J.A. (2002) The holiness of Yiddish: Who says Yiddish is holy and why? *Language Policy* 1 (2), 123–141.

Fishman, J.A. (2006) A decalogue of basic theoretical perspectives for a sociology of language and religion. In T. Omoniyi and J.A. Fishman (eds) *The Sociology of Language and Religion: Change, Conflict and Accommodation* (pp. 13–25). Basingstoke: Palgrave Macmillan.

Katz, D. (2004) *Words on Fire: The Unfinished Story of Yiddish.* New York: Basic Books.

Index